Growing Soft Fruits

Contents

2 Soil and Site
4 Tools
6 Planting
8 Protection
10 Blackberries
14 Blackcurrants
18 Cape Gooseberries
19 Chinese Gooseberries
20 Gooseberries
24 Grapes
29 Loganberries
30 Melons
34 Raspberries
38 Red and White Currants
40 Rhubarb
42 Strawberries

Soil and Site

Soft fruits will, in nearly all gardens, amply repay the time and trouble expended on them. Only if the site is subject to late frost damage or is very chalky should these fruits be excluded from the garden scheme.

Few gardeners have much of a choice of soil or site, but there are steps that can be taken to improve most soils.

Improving the Soil Structure

The main requirement of all soft fruit is plenty of organic matter. Though most soft fruits make a mass of surface roots, they also need to push down further in search of water and nutrients, and for anchorage. So the soil must be deeply broken up and capable of retaining plenty of plant food and water. At the same time, drainage must be good and there should never be an excess of water lying on the surface.

Organic matter will help improve all these conditions. It can be provided in the form of well-rotted farmyard manure, compost, spent mushroom compost or even peat, though this last is an expensive way of doing it.

Ideally, the site should be double-dug, and the organic matter incorporated in the bottom trench. Don't just dig holes for the individual plants and put a bit of manure in the bottom. This will invite water to drain into the hole, and root rotting could follow.

To double-dig, start by dividing the plot into two. Take out a trench at one end about 60cm (2ft) wide and one spade deep, across half the plot. The soil from this trench should be carted across to the other side of the plot at the same end.

Now put a liberal dressing of manure in the bottom of the trench and fork it in, breaking up the bottom of the trench to the fork's depth. The second trench is then measured to exactly the same width as the first and dug out,

Double digging

again one spade deep. The soil is thrown forward to fill the first trench. Continue in this way down the plot. When the end is reached, turn round and proceed down the other half of the plot. The soil that was carried to the end of the second half is used to fill the last trench.

Lime
All soft fruits prefer a slightly acid soil, so there should be no need to lime.

Frost
Soft-fruit blossom can be destroyed by late frosts, resulting in reduction or even complete loss of the crop. The site must therefore be carefully chosen to avoid frost pockets.

Weeds
Perennial weeds can be a real headache in soft-fruit growing, and their competition will seriously affect the crop. If there is an infestation of invasive weeds such as couch grass, ground elder or oxalis, spray the whole area with a total weedkiller containing glyphosate (Murphys 'Tumbleweed') before starting cultivations. Other perennial weeds can usually be dug out. Annual weeds should be buried during digging.

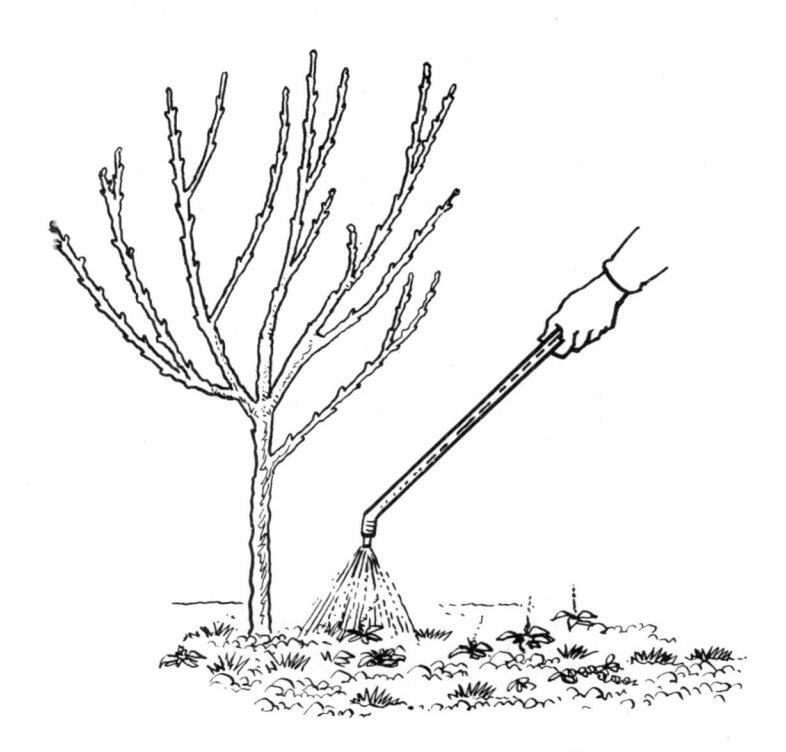

Spraying weeds with a total weedkiller

Tools

Most established gardeners will already have the tools needed for soft-fruit growing, but for those just starting, it is worth buying the right tools for the job.

Bear in mind that buying cheap tools is a false economy. Most good gardening tools, if well looked after, should last a lifetime, while cheap, badly made ones may last only one season.

Spade
A good spade is essential for the initial preparation and for planting new bushes later. If your budget can stand it, a stainless steel blade will make life a lot easier and digging a positive pleasure.

Fork
A square-tined digging fork will be used each and every season. Stainless steel here is a bit of a luxury and certainly not essential.

Trowel
A trowel may be necessary for planting strawberries and melons etc, but buy an expensive one only if you are well organized and tidy minded, as they tend to get lost.

Secateurs
A good pair of secateurs is essential. Here again, it's worth buying the best. Cheap models tend to twist when cutting thick wood. This eventually results in tearing rather than cutting cleanly, and can provide an entry for disease.

The anvil type is ideal for soft fruit. If the parrot-bill type is preferred, make sure that the blades have an adjustment to tighten them when they become worn.

Sprayer
The soft-fruit grower who can get away without spraying is very lucky indeed. There are bound to be pests and diseases even in the best-run garden and a lot of work can be saved by spraying against weeds. There are several good plastic pressure sprayers available now, and they are not expensive.

Watering Equipment
Marked increases in yields can be obtained by applying water at the right time, so some means of applying it automatically will be a boon. Ideally, use a few lengths of seep-hose that will gently apply water to the soil surface. At a pinch, the lawn sprinkler can be used. Naturally, a hose will also be needed. Here again, it is wise to stick to the more expensive kinds.

Cloches
Some soft fruits, like melons, and particularly strawberries, are especially suited to being grown under cloches. They will produce earlier yields when the price of fruit is high, and they will extend the season at both ends. Plastic tunnel cloches are ideal for strawberries and are not expensive.

The higher, barn-type glass cloches are valuable for growing taller fruit crops and are particularly useful for growing dessert grapes outdoors. They have the disadvantage that they are easily broken and expensive to replace, but you can buy sheet plastic which will last for many years.

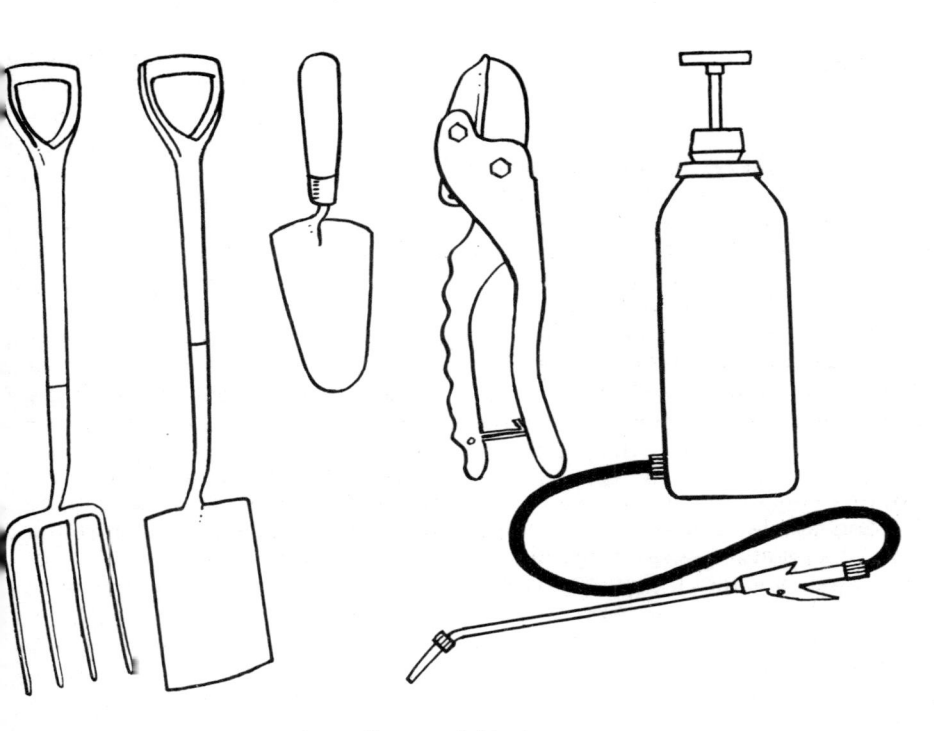

Some useful tools

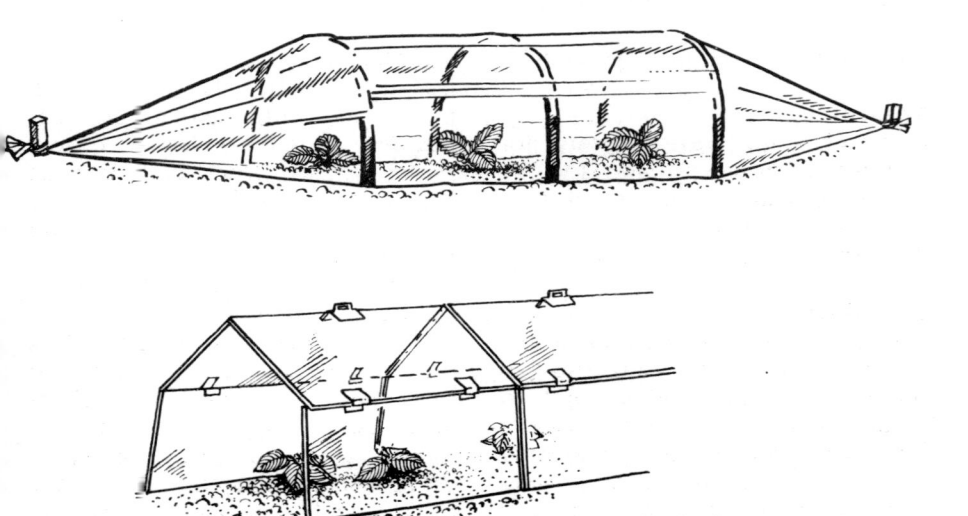

(above) Plastic and (below) glass cloches

Planting

With the exception of strawberries, melons and rhubarb, planting methods are much the same for all bush and cane fruit. Distances apart differ, though, and they are dealt with under each heading.

Most soft fruit can now be bought from the garden centre in containers. This means that they can be planted at any time of the year. It is cheaper and sometimes better, however, to buy bare-rooted plants from a specialist. Specialist fruit growers tend to carry a wider range of varieties, and their quality is often better. In this case, planting should be carried out during the dormant season from November to March. It is best to plant as early as possible after leaf-fall, while the soil still has some warmth in it. This will get the plants off to a good start in the spring.

Planting should be delayed, however, if the soil is frozen or very wet. The rule is: if the soil sticks to your boots, keep off it.

If plants arrive when the soil is frozen, leave them in their wrappings and put them in a frost-proof shed or garage until the soil thaws. If the ground is too wet, the plants can be heeled in, in a shallow trench in a drier corner of the garden, until conditions improve.

Before planting, mark out the sites for each plant. This can be done with a garden line and a few canes. Holes should be dug in the previously prepared soil, roughly to the required depth. Remember that blackcurrants must go a little deeper than they were grown at the nursery.

Set the plant in the middle of the hole and put a cane across it to make it easier to judge whether the plant needs raising or lowering, and adjust it accordingly. Spread out the roots to their fullest extent. Before refilling, scatter a handful of bone-meal over the heap of soil you have dug out. Now work the best of the topsoil around the roots. The hole should be refilled in stages, treading with your heel to ensure that the roots are really firm. After planting, apply a generous mulch of manure or compost around the plants, but not actually touching them.

Container-grown bushes are planted in much the same way. First of all, make sure that the root ball is well soaked before removing the pot. This will prevent the young roots sticking to the side of the pot, and becoming ripped off when it is removed.

Remove the container by splitting the bag down the side with a sharp knife or a pair of scissors, making sure that you don't cut the roots.

Set the plant at the right level in the hole, ie so that the top of the root ball is only just covered with soil.

The soil around the roots must be firmed, but take care not to damage the roots; always tread *round* the root ball and never on top of it.

Container-grown plants can be set at any time of the year, but if they are put in during the spring or summer, care must be taken to ensure that they don't run short of water. If necessary, give them a

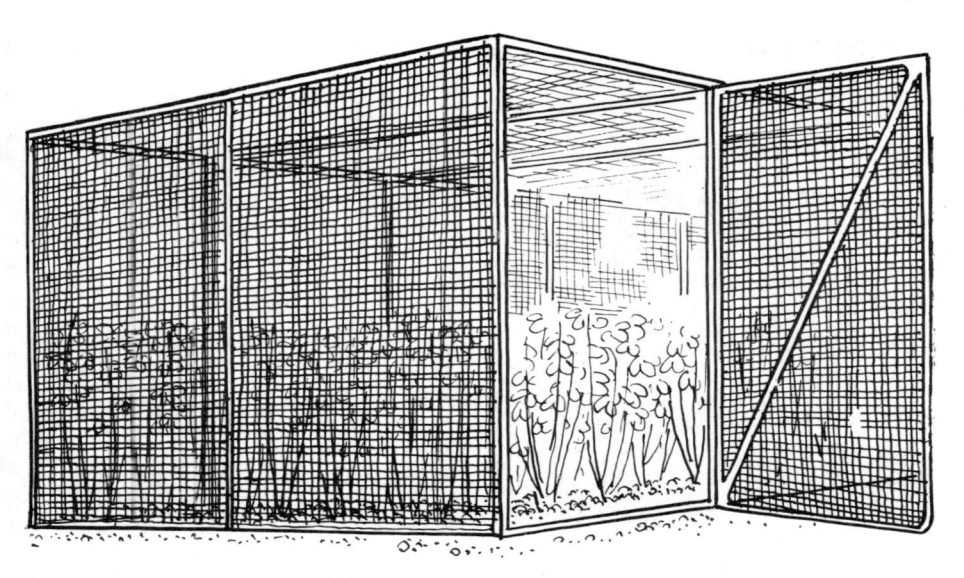

A cage will protect fruit from birds

good soaking from time to time and mulch around them with manure or compost to reduce evaporation.

Protection against Birds

Birds are a menace often at two periods during the growing season. Some birds, especially bullfinches, will eat the swelling buds of some bushes. Blackcurrants are particularly prone to attack. All types of birds will also try to strip the fruit off the bushes before you do!

Buds can be protected by spraying with a bird repellent and these are generally quite effective. Ripening fruit is best covered in some way. Low-growing crops such as strawberries can easily be covered with bird netting, making sure that the edges are well weighted down with stones or soil. The netting is easily pushed back for picking.

Taller fruit is best grown under a fruit cage. Proprietary makes can be used, though they are fairly expensive. They consist generally of a framework of metal tubes covered in plastic netting.

Alternatively, a fair amount of money can be saved by making your own. Rustic poles or angle-iron can be used as the supports. Make them 2.5m (8ft 6in) high and sink them 60cm (2ft) in the ground, at about 3m (10ft) intervals. Before setting the poles, treat the ends with a copper-based preservative to protect them.

Run thick covered wire (ex-army telephone wire is ideal) between the posts at the tops and cover the whole structure with bird netting. The netting can either be pegged into the ground with metal tent-pegs, or it can be buried in a shallow trench. If you are using angle-iron posts, put a jam jar on the top of each post to protect the netting.

Protection

Garden produce is prone to attack from many quarters, and soft fruit is no exception. Insect pests and fungus diseases are not too difficult to control by spraying, but frost can prove more difficult.

Frost
Late frosts cause damage mainly to blossom—and, of course, without blossom there can be no fruit. If frost threatens, bushes can be protected by covering with straw or even netting, though this is a tiresome chore. The material must also be removed when the danger of frost is past, and perhaps even replaced and removed several times while the threat persists.

Some fruits, notably strawberries, will do well under cloches which should give them all the protection they need. Mulches should not be applied around the plants until all danger of frost has passed. During the day, the soil will absorb some heat, which will be lost at night, thus slightly raising the temperature around the plants.

Another method of protection when frost threatens is to spray plants with water. This will actually raise temperatures at night and could be all that is needed to prevent damage.

In colder areas, all these measures may be to no avail. The only alternative then is to choose varieties that flower later and so escape the frost.

Windbreaks
Experiments at fruit research stations have shown that yields can be greatly increased by protecting fruit against strong winds. Wind can damage branches and destroy blossom and young fruits. Areas that are subject to strong winds also tend to be colder, and this can have two effects: it delays cropping slightly, but, perhaps more important, it will discourage pollinating insects.

Protection against wind also makes working outside that much more pleasant. To the serious grower, this may not seem important. But the job of pruning, for example, can be so unpleasant in high winds on a cold winter's day, that it tends to get left.

No attempt should be made, however, to stop wind completely. A solid barrier can even have an adverse effect: the wind will 'jump over' the barrier and descend rapidly on the other side at an even greater speed.

The answer is to reduce the speed of the wind by erecting a permeable barrier which filters the wind and slows it down to an acceptable level.

Evergreen hedges make an ideal windbreak in the garden since they also serve to mark boundaries and provide privacy. And, of course, they look attractive at the same time. Conifers are ideal. The fastest-growing hardy conifer for hedging is *Cupressocyparis leylandii* which, after the first year, can grow as fast as 1m (3ft) a year. Two other fast growers ideal for the purpose are *Chamaecyparis lawsoniana* and *Thuja plicata*.

One disadvantage of having hedging round the fruit garden is that the hedge will provide a home for insect pests, so make sure that when the fruit is sprayed, the hedge is treated as well.

An alternative to hedging is to erect a plastic barrier of special windbreak material. This can be obtained at most garden centres or shops, and, though not cheap, will last well over ten years. It should be erected on stout posts on the windward side of the fruit garden. When erecting the windbreak, it should be borne in mind that protection will be afforded over a distance up to thirty times the height of the barrier. Protection naturally decreases further away from the screen, however, and the area adequately protected is about ten times the height of the barrier. Thus a screen 1m (3ft) high will give adequate protection to bushes planted up to 10m (30ft) away from it. It is wise to plant low-growing crops to the leeward side of taller crops so that they too are protected.

Windbreaks can have one serious consequence that must be avoided at all costs. They can create their own frost-pocket, by trapping moving air-frost that would otherwise have escaped. If the garden is on a site that is subject to air frosts, and especially on sloping ground, the windbreak should be raised 30cm (1ft) or so off the ground to allow trapped frost to escape.

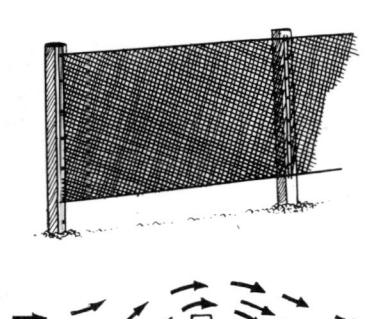

(above) A plastic barrier makes an effective windbreak, giving protection over a distance up to thirty times the height of the barrier; (centre) a solid barrier can have an adverse effect, as the wind will descend rapidly on the far side; (below) evergreen hedges make ideal windbreaks

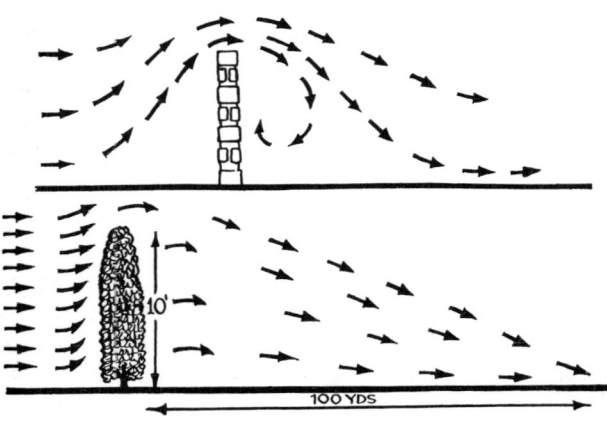

Blackberries

Blackberries are fairly rampant growers, though some varieties are less vigorous. Cultivated varieties will produce larger fruits than the wild types, though the flavour is considered by some to be inferior.

Though they grow vigorously, they can be trained against a fence or wall, or on posts and wires on the perimeter of the fruit garden, and thus still be valuable even in the small garden. They are tolerant of most soils and have the advantage that they will produce good crops in semi-shade. Cold winds in winter can damage the canes, so they should not be grown in a very exposed situation.

Planting

Plant in the dormant season between October and March, preferably before Christmas when the soil is warmer. They should be at least 2m (6ft) apart, especially the more vigorous varieties. Do not plant if the ground is too wet or frozen.

After planting, cut the shoots back to three or four buds above the ground

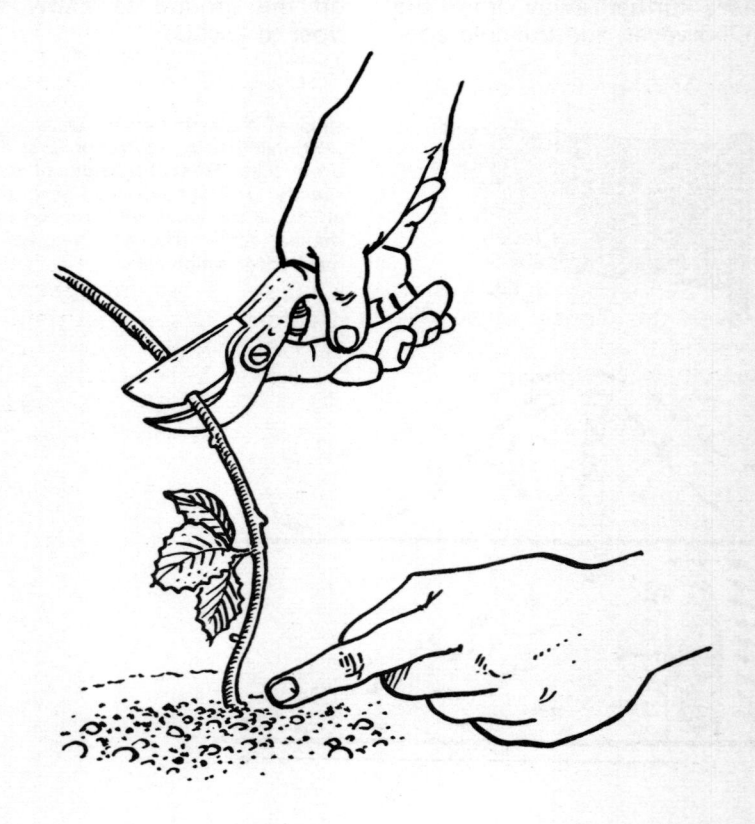

If the soil is likely to dry out, give the plants a good mulch of manure or straw. Then cut the shoots back to 3–4 buds above ground level.

Cultivation

Keep the weeds down at all times. This is helped by mulching with well-rotted manure or compost, which also helps to retain water. They need little feeding, and on good soils will need only a little nitrogen in the spring. Use sulphate of ammonia at about 56g per sq m (2oz per sq yd). If farmyard manure is not available, apply about 56g per sq m (2oz per sq yd) of sulphate of potash annually and the same amount of superphosphate every three years.

Never allow the plants to suffer from lack of water. They will benefit from irrigation, particularly when the fruit begins to swell.

Harvesting

Pick the fruit when it is just ripe. This is normally in July, August and into September. It is best to pick over the canes every few days, rather than leave them until the majority have ripened.

Propagation

Propagate by tip layers. Bend down the tips of shoots and bury them about 15cm (6in) deep. They will quickly root and can be removed in the spring.

Pests and Diseases

Raspberry beetle grubs attack the fruits and there is sometimes some tip damage as well. Control by spraying with malathion or fenitrothion when the first flowers open and again at petal fall. The pest is widespread and not seen until too late, so precautionary sprays are recommended.

Blackberry purple blotch causes purple areas on old stems. It should be controlled by spraying with benomyl in mid-June and again a month later.

Botrytis is a fungus disease which is seen as a grey mould on fruit and can cause the canes to turn milky white in winter. Spray with benomyl when the first fruits turn pink. If the season is wet, spray again just before picking.

Virus diseases

Blackberries are subject to virus diseases which cause a stunting of growth and a lowering of yield. There is no cure. Plants should be removed and burnt, and healthy stock planted in a different part of the garden.

Varieties

Merton Early, an early variety of medium growth and good flavour.

Bedford Giant, early to mid-season and of medium vigour. Fruits are very large but flavour is only fair.

Himalaya Giant, a mid-season variety, very vigorous and thorny. Flavour is quite good.

Oregon Thornless, a mid-season variety, not vigorous and with fruits of good flavour.

Blackberries

Training
Young plants can be set adjacent to a fence or wall, or a special support system of stakes and wires can be erected. The stakes should be about 1.5–2m (5–6ft) out of the ground, with four wires about 30cm (1ft) apart spaced from the top. The posts should be about 6m (20ft) apart.

As the canes grow, they should be tied in to the wires, spacing them so that they receive a maximum amount of sun and air. Spacing also helps prevent the spread of disease.

The fruit is borne on short spurs that grow out of last year's wood. After fruiting, the fruited wood must be cut out. So, in order to maintain a supply of two-year-old fruiting wood, it is necessary to keep young and old wood separate. There are three ways of doing this.

The Weaving System
Here the older growths are woven in and out of the first three wires, while the current season's canes are taken up and tied along the top wire. Once the older wood has fruited, the canes are cut back hard, and the new canes woven in to take their place. This is a good, labour-saving method since little tying-in is required.

The Fan System
The fruiting canes are tied in to a fan shape. The centre is kept open and the young canes tied in and along the top wire. This method looks tidier and allows more even spacing, but is quite time consuming.

The Rope System
The fruiting canes are tied on one side of the stool and the young canes on the other. It needs less labour, but bunches the canes together.

Different methods of training blackberries

WEAVING

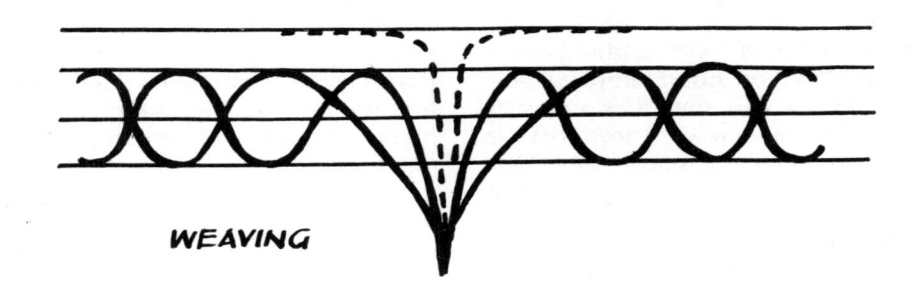

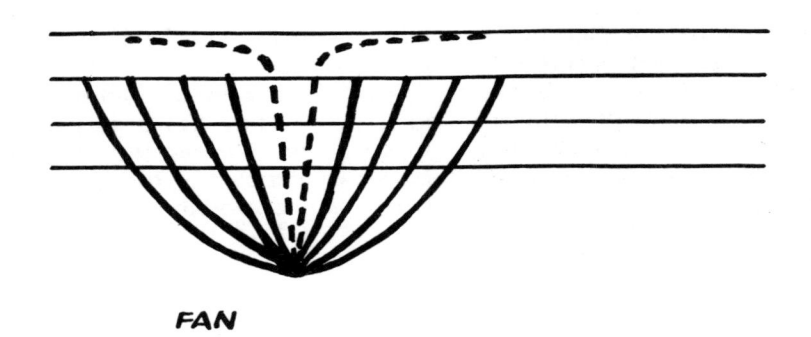

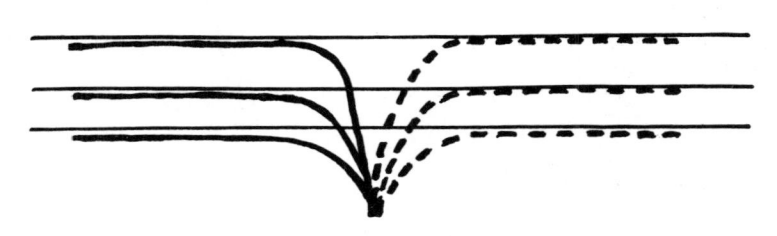

FAN

ROPE TRAINED, ONE WAY

Blackcurrants

Blackcurrants are one of the most important soft fruits in the garden. They produce good crops almost unfailingly and the fruits are very high in Vitamin C. They prefer a sunny, open position, but will tolerate partial shade.

Planting

Plant during the dormant season between November and March, preferably before the end of the year.

Blackcurrants should be grown on a 'stool', ie they should be planted slightly deeper than in the nursery to encourage shoots to grow from below ground. On no account should a length of stem be left above ground so that they grow on a 'leg', as is the practice with redcurrants and gooseberries.

They need as much light and air as possible, so allow at least 1.5m (5ft) between plants and 1.8m (6ft) between rows.

Pruning

Blackcurrants produce their best fruit on two-year-old wood, so the object of pruning is to produce as much young wood as possible. After planting, cut all the stems back hard to within about 2.5cm (1in) of the ground. The plants will not fruit in the first year. At the end of the year, cut out the weakest shoots and leave the strongest to fruit.

After that, pruning consists of cutting out the wood that has borne fruit that year and anything that looks too weak.

Pruning should be carried out just after the fruit has been picked to give the bushes the longest possible chance to produce more strong young growth.

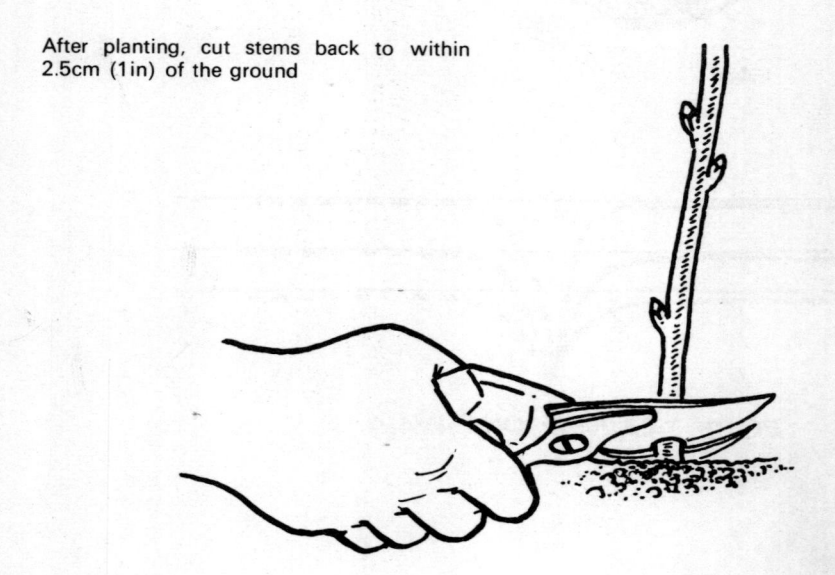

After planting, cut stems back to within 2.5cm (1in) of the ground

Propagation

Blackcurrants are quite easily propagated from hardwood cuttings taken during the autumn. Select healthy shoots from the current year's growth and remove the unripened tips. Trim the shoots to about 20cm (8in) long, cutting the base just below a bud.

Unlike most cuttings, the buds should not be removed, since these will produce stems from underground, making the stool.

Cut a shallow trench in a corner of the garden in a sheltered position and line the bottom with a little sharp sand. The cuttings are then inserted in the trench about 15cm (6in) apart so that just two or three buds show above ground. Firm the cuttings in well with your heel.

Leave the cuttings there for a year. In the following autumn, many will have rooted, and these can be transferred to their permanent positions.

Propagating blackcurrants from cuttings

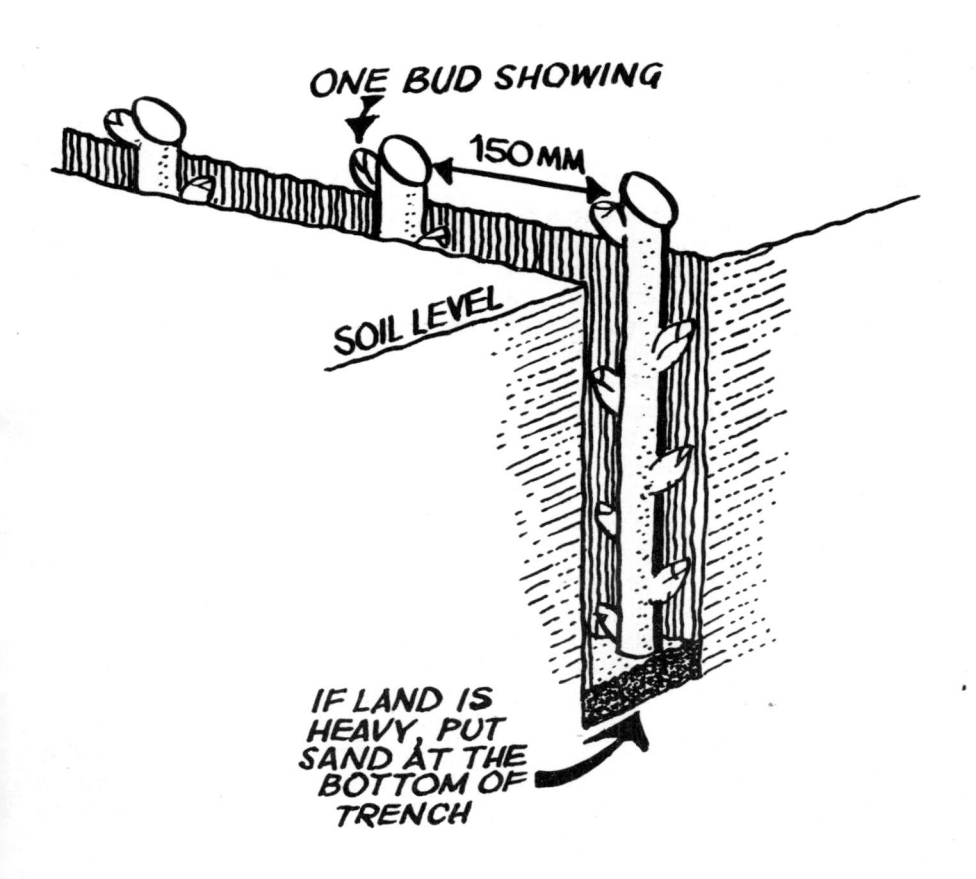

Blackcurrants

Feeding
Blackcurrants are greedy plants and respond well to feeding and watering. If quantities of manure are available, they will thrive on liberal applications applied at any time. Alternatively, mulch them well with well-rotted compost or spent mushroom manure.

The nutrients from the mulches should be supplemented with a feed of sulphate of ammonia applied at 28g per sq m (1oz per sq yd) in March.

Cultivation
Keep the plants weed-free at all times. This should not be done by deep hoeing since they will produce a mass of surface roots. It is better to keep the weeds down by spraying with a herbicide such as 'Weedol' for annual weeds or 'Tumbleweed' if perennials are present. The mulching during the season will also help weed control.

Blackcurrants will benefit from watering, especially when the fruit is swelling. In any case, the bushes should never be allowed to dry out.

Harvesting
Pick the fruit when it is ripe and use it straight from the tree. It can be used raw or cooked, or for jams, preserves or wine. Fruits can also be frozen.

Pests and Diseases
Blackcurrants, like many other soft fruits, are covered by a Ministry certification scheme. They are

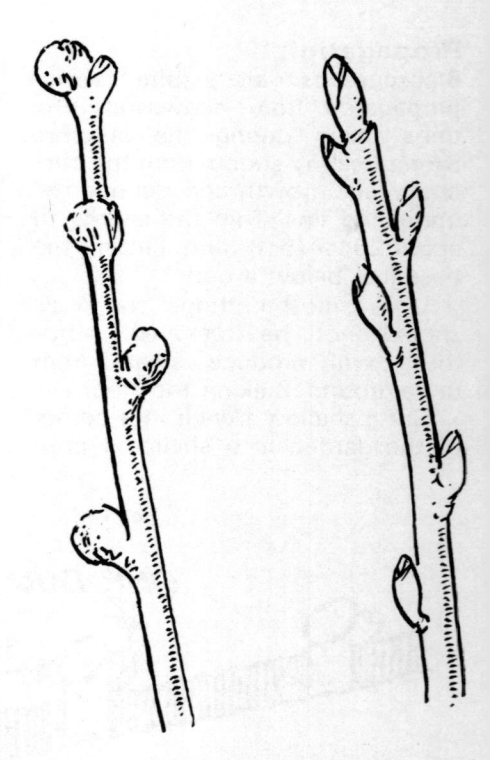

(left) Swollen buds; (right) normal buds

prone to a virus disease known as 'reversion', which is transmitted by blackcurrant gall mite. The mites destroy the buds, and the virus results in rapid loss of vigour and a marked reduction of crop. It is essential, therefore, to buy only plants that are being grown for certification. Since reversion is very difficult to detect in one-year-old bushes, certificates are issued only for plants between two and five years old. But if one-year-old bushes have been entered for certification it is fairly safe to buy them. Hand-picking any swollen buds may limit the disease to some extent, but because of its seriousness, the bushes are best removed and burnt.

Aphids are one of the most common pests of all fruit. They cluster round the growing tips sucking the sap and causing the leaves to become puckered and distorted. Fortunately, they are not difficult to control, but spraying should be carried out as soon as they are seen. Even if they are not visible, the bushes should be sprayed as a precaution, using dimethoate.

Blackcurrant leaf midge causes much the same symptoms as aphids, and is controlled by the same spray.

Sawfly sometimes attacks, stripping and skeletonizing leaves. As soon as the symptoms occur, spray with fenitrothion.

Leaf spot is a disease that causes dark spots on the leaves. These increase to become dark blotches. Control with triforine applied when the fruit is small and again three weeks later. Spray again after cropping and three weeks later.

Mildew will sometimes attack blackcurrants, though it is more common on gooseberries. It produces a white, powdery coating on leaves, shoots and fruits. It too can be controlled with triforine applied at the same time as for leaf-spot.

Varieties

If varieties are carefully selected, it is possible to have fresh fruit from the end of June to early August. So, it is well worth while planting a selection of varieties to provide a succession.

Boskoop Giant is one of the favourites in the garden because it is early. It crops well, but, because the skins are thin, it does not last long on the bush.

Laxton's Giant is possibly better than Boskoop because the skins are thicker, and it crops at about the same time. It produces good crops of large, well-flavoured fruit.

Blackdown is a new variety classed as a second early. It produces heavy crops of large, sweet berries and is resistant to mildew.

Seabrooks Black is a mid-season variety that forms a large bush. It crops heavily generally but is not always reliable. The medium-to-large berries are somewhat acid and thick-skinned.

Wellington XXX (Wellington Triple Cross) is another mid-season variety that forms a very large bush. It needs more space than most. It is a heavy and regular cropper with medium berries of good flavour.

Baldwin is an old variety cropping mid to late season. It produces medium-sized bushes and crops heavily. The berries have a slightly acid flavour and it is especially recommended for freezing.

Amos Black is a late variety. It also flowers very late, so could miss spring frosts. It is a bit variable in cropping and the fruit tends to fall.

Malling Jet is a new introduction. It too fruits very late in the season and crops heavily in most seasons. Fruit is medium-sized and of good flavour.

Cape Gooseberries

The Cape gooseberry is tender and should therefore only be grown outside in warmer areas, or in the greenhouse. It is generally grown as an annual.

Planting
It is best to grow these plants in soil that is not too rich, or they will make too much vegetative growth at the expense of fruit.

They are best raised in the greenhouse from seed sown in March. When the seedlings are large enough to handle, prick them out into small pots. They should not be planted outside until the end of May when all danger of frost has passed. Plant them 75cm (2ft 6in) apart with 100cm (3ft 6in) between rows.

For greenhouse growing, pot them on into 23cm (9in) pots.

Cultivation
Outside, keep the plants weed-free and well watered. Inside, spray them over with water regularly, and tap the plants from time to time when they are flowering to ensure pollination.

Once the fruits have set, they should be fed weekly with a proprietary tomato fertilizer.

Pruning
When the plants are 30cm (1ft) high, pinch out the growing tip to make them bushy. Regular pinching out of sideshoots is necessary to ensure that the size of the plant is restricted. Support the plants with a cane.

Harvesting
The fruits are ready to pick when the calyx changes from green to brown and has a papery texture. Inside the calyx the fruits will be a golden-brown colour. Pick them with the calyx still on.

They may be eaten raw from the plant or they can be stored for some time in dry, cool, but frost-free conditions.

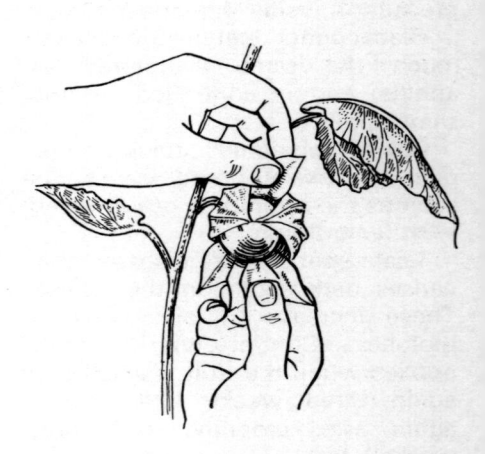

Cape gooseberry

Pests and Diseases
Cape gooseberries are generally trouble-free, though they can be attacked by whitefly and aphids. Whitefly are particularly troublesome under glass, where they suck the sap, causing distortion of the leaves and a slowing down of growth.

They can be controlled by spraying with resmethrin as soon as the pests are seen and again at seven-day intervals. Aphids are easily controlled by the same spray.

Chinese Gooseberries

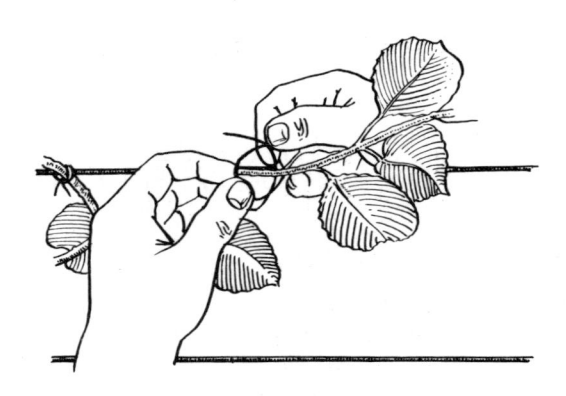

The Chinese gooseberry is often grown as an ornamental plant, though the fruit is very attractive, has a good flavour and is very high in Vitamin C. It is a vigorous climber and will do well on a south-facing wall or rambling over a pergola. Here is a plant that is perhaps better grown in the ornamental garden than in the fruit garden.

Tying Chinese gooseberries

Planting
Though tolerant of most soils, the Chinese gooseberry prefers a warm, moist soil. In the north, it is best grown under glass since the young growths are susceptible to damage by spring frosts.

When buying plants or raising from seed, remember that, for a crop of fruit, you will need a male and a female plant.

Plant 5–6m (17–20ft) apart, and provide a strong support system since these plants are very rampant growers.

If several plants are being set, plant one male plant to about six females to ensure good pollination of the flowers.

Pruning
In the early stages, the long growths should be tied in to fill the framework or cover the wall.

When the plant is established it can be pruned back in the winter to encourage the production of fruiting spurs. This involves cutting back the previous summer's growth to three or four buds.

Propagation
Chinese gooseberries are sometimes offered by seedsmen and they can quite easily be raised in this way. The best varieties, however, are selected and are therefore best raised from cuttings.

Pests and Diseases
Chinese gooseberries are generally trouble-free, but can be attacked by whitefly and greenfly. Both these can be controlled by spraying with resmethrin. For whitefly, spray at seven-day intervals.

Varieties
There is no variety especially bred for fruit though the plant (*Actinidia chinensis*) is sold sometimes as 'Chinese Gooseberry' and sometimes as 'Kiwi Fruit'.

Gooseberries

Gooseberries are one of the easiest and most valuable soft fruit crops for the small garden. Even when neglected they produce a crop, though they will amply repay good cultivation. They are tolerant of most soils and conditions.

Planting
Buy good, well-shaped bushes from a reputable source, choosing two-year-old bushes if possible. They require a lot of room, since they grow large and are quite uncomfortable to tend if grown too close.

Plant in rows at least 1.5m (5ft) apart with 2.3m (7ft 6in) between rows. Gooseberries can also be grown as cordons, and these should be about 30cm (1ft) apart.

Gooseberries are grown on a 'leg', so they should be planted at the same level at which they were growing at the nursery.

Cordons should be planted against a post and wire structure and should be tied in as soon as possible after planting.

Pruning
After planting, cut back the shoots by about half, cutting to a bud. If the variety has a drooping habit, cut back to an upward-facing bud, otherwise to one that faces down.

For the first few years, the aim should be to establish a strong framework to support heavy crops. Cut back the strong leading shoots

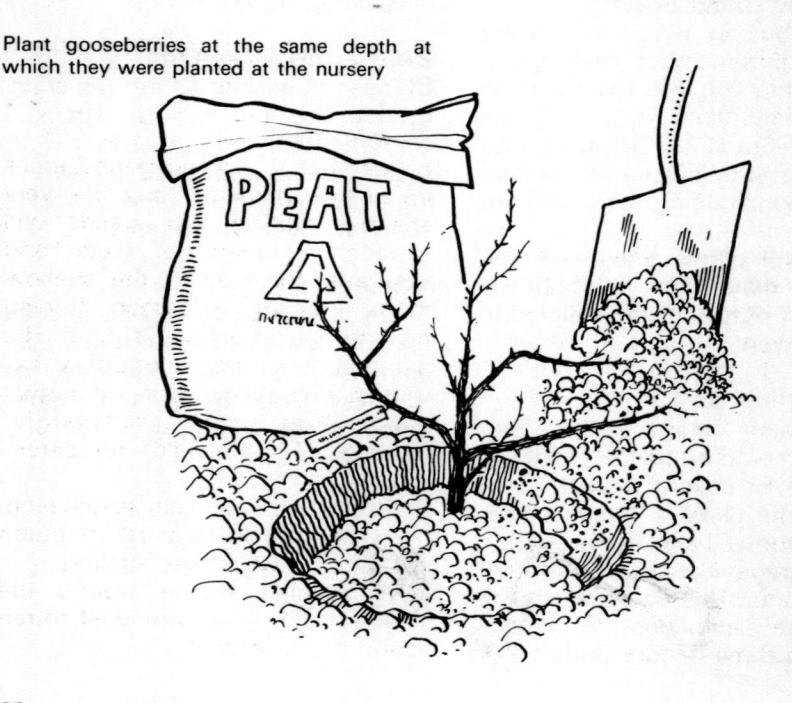

Plant gooseberries at the same depth at which they were planted at the nursery

by about halfway, and remove any that are weak or growing into the centre of the bush. Remove also branches that are crossing, overcrowded or damaged. At the same time, to encourage the formation of short fruiting spurs, cut back the side growths to about 7.5cm (3in).

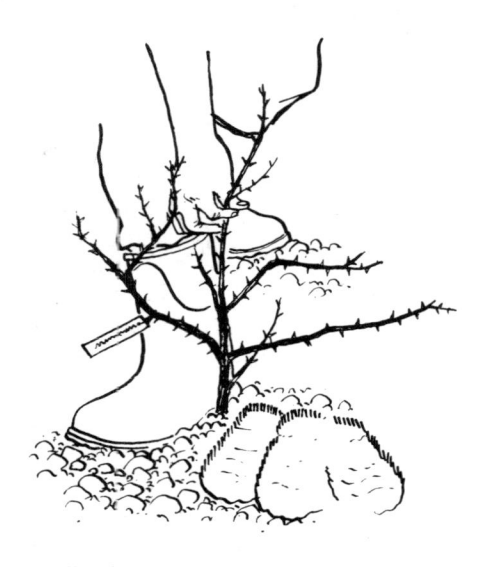

Heeling in

In subsequent years, winter pruning consists of cutting out weak and damaged branches and those that are growing into the centre.

When the shape of the bush has been formed, summer pruning can start. After about the third week in June, all sideshoots are shortened to about five leaves, to encourage the formation of fruit buds.

If plants are being grown as cordons, it is essential to prune regularly, summer and winter. From about the middle of June, all the sideshoots should be cut back to about five leaves. The leader (growing point) should be left until it is as high as is required, when it can be treated in the same way. In the winter, after leaf-fall, the shoots are further shortened back to three buds.

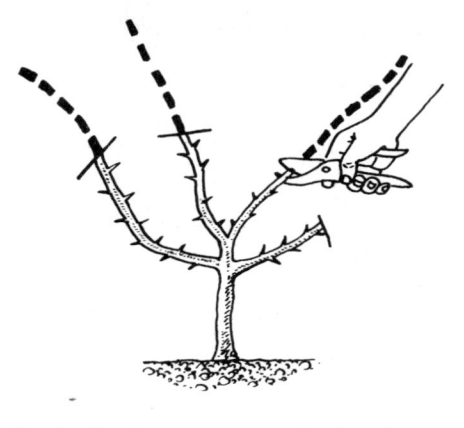

For the first few years, prune leading shoots by about half

Cultivation

It is essential to keep gooseberries weed-free because they are very shallow rooting and weeds will seriously compete for water and nutrients. Because of the shallow roots, digging or deep hoeing round the bushes should be avoided. It is best to control weeds with a weedkiller, using paraquat (Weedol) for annual weeds or glyphosate (Tumbleweed) for perennials.

Don't allow suckers to develop. They should be torn off in much the same way as rose suckers.

If possible, mulch around the bushes each year with well-rotted manure or compost.

21

Gooseberries

Feeding
Too much nitrogen will cause excessive growth at the expense of fruit. But gooseberries are very susceptible to potash deficiencies.

So, besides the dressing of manure or compost, sulphate of potash should be applied each February at a rate of 22g per sq m ($\frac{3}{4}$oz per sq yd). If manure is not available and the plants seem slow to make growth, supplement this with 28g per sq m (1oz per sq yd) of sulphate of ammonia and 22g per sq m ($\frac{3}{4}$oz per sq yd) of superphosphate.

Harvesting
The first fruits can be picked before they are ripe and used for cooking. This will give the remainder a chance to grow larger and they can be picked ripe for dessert purposes.

Propagation
Gooseberries can be increased by cuttings. Take shoots from the current season's growth in the autumn and trim them to about 30cm (1ft) long. Remove the spines and all the buds except the top three or four. The cuttings are then placed in a slit trench, as advised for blackcurrants (see p 15).

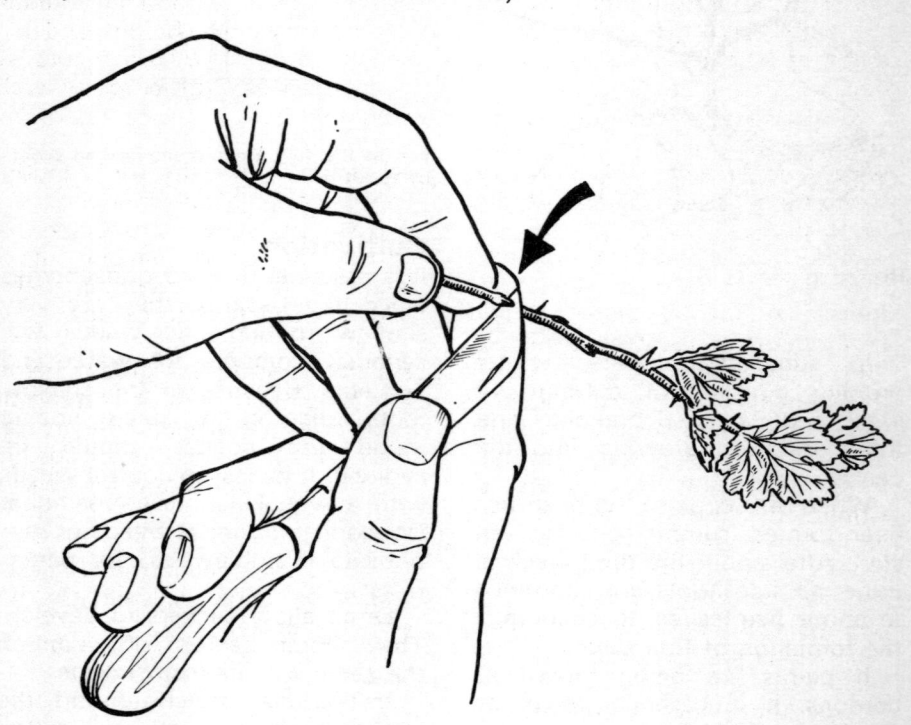

Removing a bud from a shoot prior to propagation

They should have rooted by the following autumn, when they can be planted out. For bushes, prune the main stem back to about four buds, and subsequently as previously advised. For cordons, allow the main stem to grow, and cut back all sideshoots.

Pests and Diseases

Bullfinches are one of the worst pests of gooseberries early in the year. They peck the swelling buds and can completely strip plants. There are one or two bird-repellent sprays available, but the only sure method of control is to cover the plants with netting or grow them in a fruit cage. If birds do attack, it is best to delay pruning until after the buds break in order to ensure that cutting back is done to a live bud.

Gooseberry sawfly attacks and skeletonizes leaves. Keep a close watch for damage and spray with fenitrothion as soon as the first signs are seen.

Aphids and green capsid attack the tips of shoots, leaves and fruits. Control by spraying with fenitrothion or dimethoate immediately after flowering.

American gooseberry mildew is the most common and serious disease. It covers leaves, shoots and fruits with white or brown fur. Control by spraying with triforine when the disease is first seen, and repeat two weeks later. If the second spray will not allow a suitable time to elapse before harvesting, wait until the fruit has been picked.

Leaf spot shows as dark, angular spots on the leaves. Spray with triforine when they are first seen.

Botrytis shows as a grey mould and is particularly likely to attack in wet seasons. Control by spraying with benomyl after flowering and again after cropping if the season is wet.

Varieties

A longer season of harvesting can be maintained by careful selection of varieties.

Keepsake is a green berry used for eating fresh or cooking. It is classed as a second early, cropping in June/July. The flowers are susceptible to spring frosts and the plant to mildew.

Whitesmith is a good all-rounder. It produces white berries and is a second early. It grows vigorously and produces large crops of well-flavoured fruit.

Whinhams Industry is another heavy-cropping all-rounder. It is a mid-season variety producing red berries of good flavour. It is vigorous with a drooping habit, and is susceptible to mildew.

Careless produces green berries and crops very well. It is generally used as a mid-season culinary variety, but when ripe has a good flavour eaten fresh. Good for freezing and for jam.

Leveller is a yellow, mid-season variety. It crops well and the flavour is excellent.

Grapes

It is often argued that grapes cannot be grown successfully in Britain except under glass. Nothing could be further from the truth. In fact, Britain was at one time a major wine-growing area, and grapes for wine are now making a comeback.

Preparation

Grapes are tolerant of a wide range of soils. In fact, they will fruit better on soil that is not too rich. On very fertile soils they tend to make a lot of vegetative growth at the expense of fruit. They are very deep-rooting and must be given good drainage if they are to succeed.

If they are being grown against a wall, it is a good idea to prepare the soil locally before planting.

Dig out a large hole about 1m (3ft) square and at least 45cm (1ft 6in) deep. Put a layer of brick rubble in the bottom of the hole and cover this with a good depth of well-rotted manure or compost. The soil can then be replaced and the vines planted.

If the vines are to be grown in the open ground, they will need supporting on a post-and-wire structure.

Three wires will be needed, the first 45cm (1ft 6in) from the ground, the top one 1.2m (4ft) high with one in between. Each plant should also have a stout stake adjacent to it.

Preparing the ground for planting vines

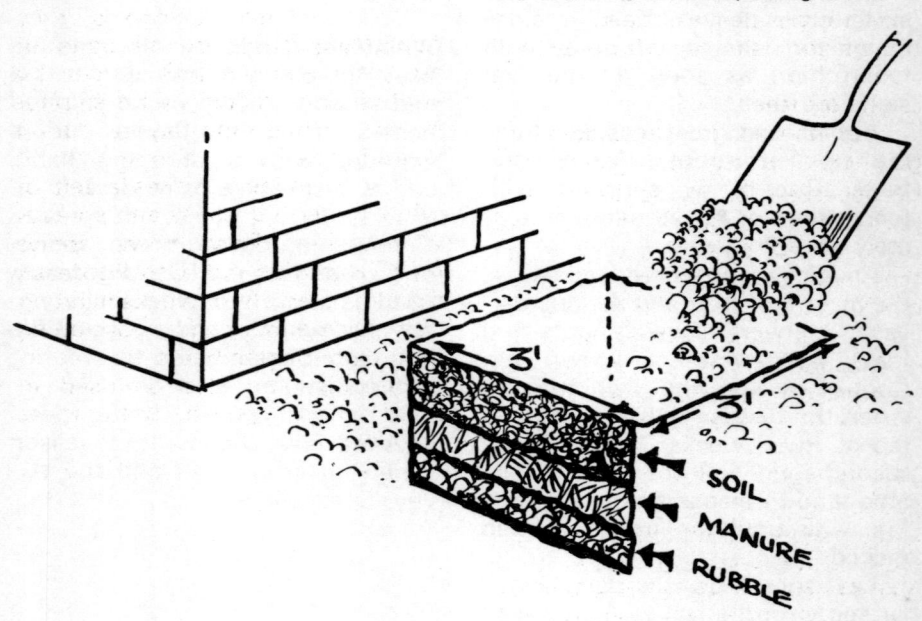

Planting
Plant in early October. Plants will generally be bought in pots. Water the pot well before removing it. Set the plant at the level it was in the pot, and surround it with the best of the topsoil. Firm well and water.

If the vines are being planted against a wall, they should be set at least 25cm (10in) from it.

Plant 1.2m (4ft) apart, and after planting, prune down to leave no more than three buds. See overleaf for pruning methods.

Growing Under Cloches
Under tall barn-type cloches, grapes have a better chance of ripening, so some of the later varieties can be grown. The same system of pruning is used, with the cloches covering the fruiting cane and the replacement canes being grown up a bamboo cane through the space in the top of the cloche.

Cultivation
Grapes should not be allowed to go short of water, so irrigation may be necessary in dry spells. This is particularly important if they are being grown against a wall, since they will dry out much more quickly, and often, in this position, do not get the benefit of rain.

A good mulch of well-rotted manure or compost each spring will also help prevent water loss.

Thinning
Grapes grown for wine do not really need thinning, but if they are required for dessert purposes, they will. Otherwise the berries will be very small. The work should be done with a special pair of vine scissors.

Start when the biggest berries are about the size of a pea. Go over the bunch, removing any that are seedless, badly placed or that spoil the shape of the bunch. Come back as the grapes swell further and thin to allow the best berries room without touching each other.

Protection
Once grapes start to swell, they are very prone to attack by birds and wasps. Net the vines against birds, and attract wasps away with containers of beer or sweetened water.

Pests and Diseases
Generally, outdoor grapes are free from pests. They do, however, often succumb to attacks by two different mildews.

Powdery mildew makes grey powdery patches on all parts of the plant and the fruit. It can be controlled by dusting with sulphur in early June and again during flowering.

Downy mildew makes a felt of white under the leaves and spreads very rapidly. Preventative sprays *must* be carried out. Use Bordeaux mixture every two weeks starting after flowering and continuing until early September.

Benomyl can also be used to good effect against both these mildews.

Grapes

Pruning
There are several methods of pruning, but the simplest and best for the home gardener is known as the Single Guyot system.

Single Guyot System
When the three buds begin to grow, remove the weakest one and allow the other two to grow. These are trained up the stake. After a few weeks, it will become obvious which is the stronger of the two. Cut the weakest down to three buds.

Throughout the season, nip off all the lateral buds to within 2.5cm (1in) of the main stem. In January, this stem too is pruned down to three buds.

In the spring, the buds will start to grow again. Remove all but the two strongest shoots once more. This time, both are trained to the stake, one to provide a fruiting cane the following year, and the other to form a replacement cane.

Thus the pattern for future pruning is set. Each year, a fruiting cane and a replacement cane are grown.

Again, during the season, pinch out all the lateral growths to within 2.5cm (1in) of the main stem.

In January, lower the stronger of the two canes and tie it to the bottom wire. It can then be cut back to about eight buds. The other cane is cut hard back again to two or three buds.

In the following year and in succeeding years, three canes are allowed to grow instead of two: that is the one fruiting cane tied to the wire, and the two replacement canes tied to the stake.

Once again, the sideshoots should be pinched back to 2.5cm (1in) on the replacement canes, but they should be allowed to grow on the fruiting cane.

Once the flower trusses show, select about two or three of the strongest laterals that are well spaced along the stem, and tie these in to the wires as they grow. The remainder are pinched out. In subsequent years, more laterals may be allowed to grow, but never more than eight to ten. Overcropping will result in poor crops the following year.

When the laterals reach the top wire, they should be pruned to about three leaves above it. Pinch out all the sub-laterals.

In November, as soon as possible after leaf-fall, cut out the cane that has fruited. The strongest of the two replacement canes is then lowered and tied in to the bottom wire to form the next year's fruiting cane. It should be cut back to about eight buds as before. The weaker of the two canes is cut back to three buds, and the process begins again.

Double Guyot System
This is much the same as the single system, except that a shorter fruiting cane is trained either side of the main stem. This is possibly better for wall training since it looks neater.

In this system, because two fruiting canes are needed, three replacement canes must be grown each year. There is no need to grow them quite so tall, however, 90cm (3ft) being quite enough.

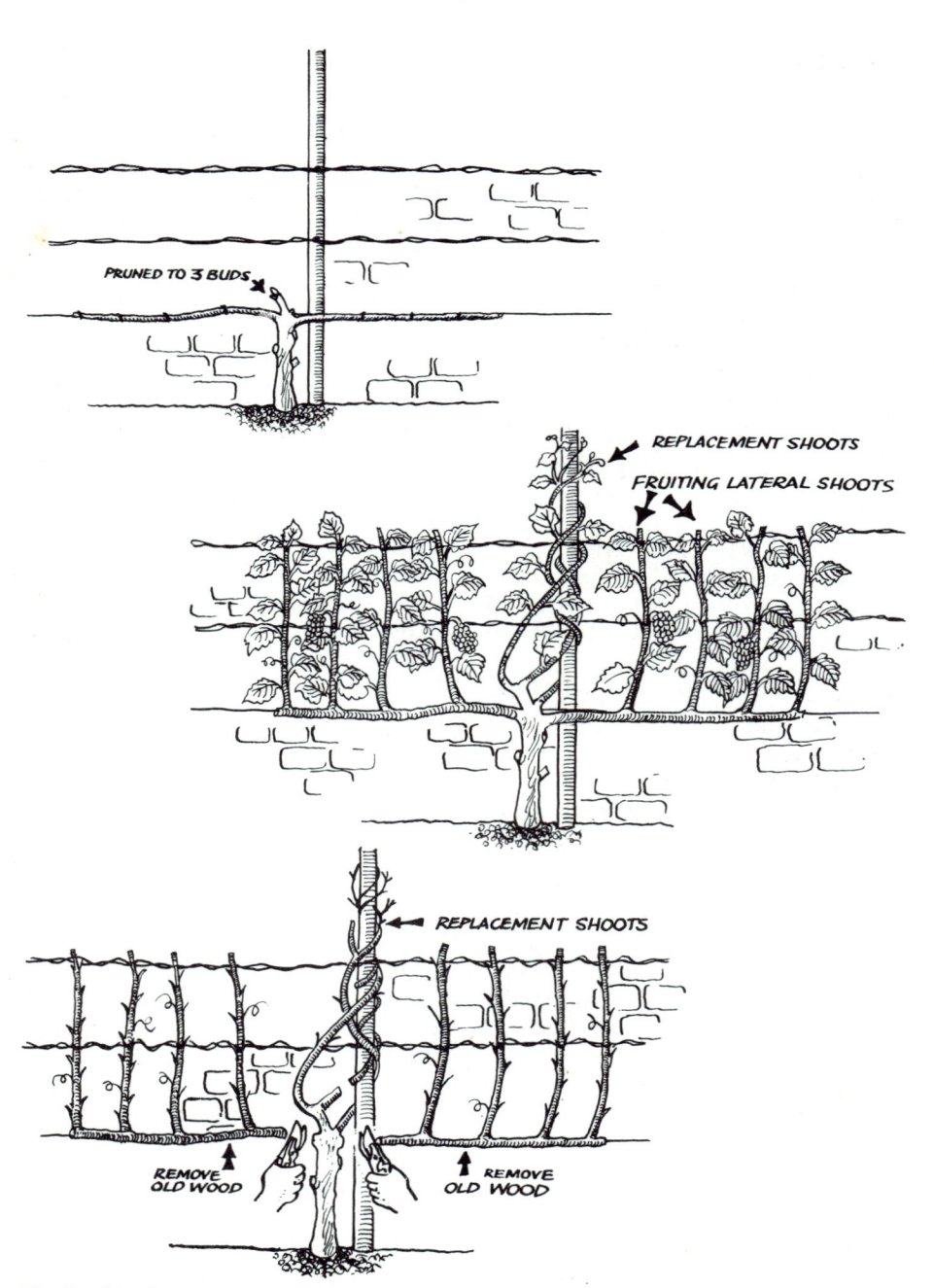

The Double Guyot System is similar to the Single Guyot System, except that two fruiting canes are trained, one either side of the main stem, as shown

Grapes

Varieties

Muller Thurgau. An early ripening variety used widely for wine production. It grows strongly and crops heavily, but is unfortunately subject to disease attack, especially mildew and botrytis.

Brant. A Canadian variety cropping early and so suitable for outside growing. It produces small black berries of quite good flavour. It is especially suitable for growing on walls and is noted for its autumn leaf colour.

Cascade. A very early French variety producing small black grapes suitable for wine. It is vigorous and heavy cropping, but susceptible to mildew.

Chasselas Dore de Fontainbleau. An early variety especially recommended for growing on walls. Sometimes also known as 'Royal Muscadine', it produces large, greenish yellow berries.

Madeleine Angevine 7972. A fairly vigorous grower producing white berries for wine making. It is especially recommended for unfavourable conditions.

Madeleine Sylvaner 28/51. An early variety especially good under cold conditions. It produces regular crops of well-flavoured grapes and makes excellent wine.

Wrotham Pinot. An excellent outdoor black grape. It is vigorous and heavy cropping and produces berries of fine flavour and sweetness.

Muscat Precoce de Saumur. A very early, hardy variety that produces good crops of sweet, well-flavoured white berries.

Loganberries

The loganberry was first raised in America in 1881. Its popularity declined when it was affected by virus and plants lost vigour. Now, however, more vigorous strains have been selected and bred for freedom from virus, so it is once more a valuable crop for the small garden. It has a very rich flavour and is particularly good when preserved.

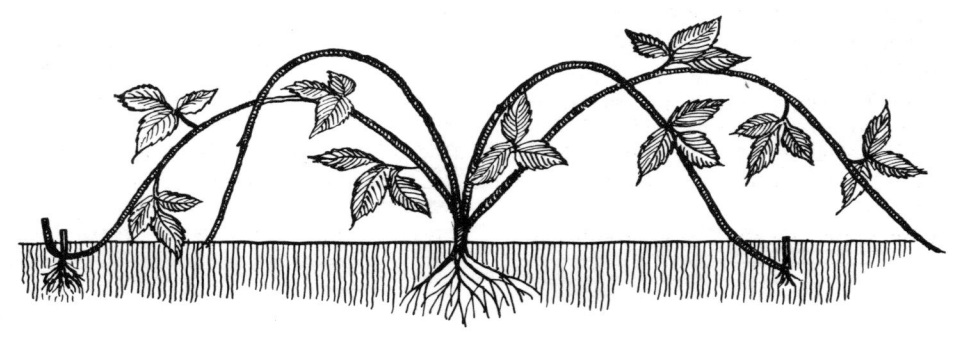

Propagation is carried out by tip layering, as shown, in the same manner as for blackberries (page 11).

Cultivation
Loganberries are cultivated in exactly the same way as blackberries (see page 11).

Pests and Diseases
Unfortunately, loganberries suffer the same pests and diseases as blackberries, plus a few more:

Spur blight. This disease causes purple blotches around the spurs. They become silvery in winter. Control by spraying with dichlorfluanid starting early in the season and continuing at 14-day intervals until the first pink fruits show. Then the spray should be changed to benomyl. Change back again after picking.

Cane spot. This causes small, dark sunken spots on the canes. The control for spur blight will be sufficient to control it also.

Varieties
It is important to buy healthy stock that is guaranteed free from virus. The Ministry of Agriculture has included the loganberry in its certification scheme since 1959, and only plants carrying a certificate of purity, health and vigour should be bought.

Loganberry. The thorned loganberry derives from a Ministry-raised stock—LY 59. It is of medium vigour and good flavour.

Thornless Loganberry. This, when first introduced, was considered to be less vigorous and of poorer flavour. Now a vigorous, well-flavoured stock has been selected. It is known by the number L654. It is of medium vigour, heavy cropping and of good flavour.

Melons

Melons can be grown successfully in a heated greenhouse, a cold greenhouse, in frames, under cloches, or, in warm, sunny and sheltered gardens, in the open. Newer varieties have been developed that will withstand lower temperatures, though they have lost nothing of their flavour and delicacy.

Melons are grown as annuals, raising them from seed each year. Though some varieties can be grown on outside, they must all be raised in heat. Direct-sown in the open, it is unlikely that the fruit would have time to come to maturity.

Sow melon seeds in moistened soil-less compost

Sowing

Ideally, sow the seeds in a greenhouse where a little bottom heat can be given. They can, however, be quite successfully raised on the windowsill, provided a little care is taken. They must, for example, be brought into the room at night, and never left behind the drawn curtains where the temperature is bound to be lower.

Sow the seed in April for planting outside in early June when all danger of frost has passed.

They are best sown in moistened soilless compost, placing the seeds on their sides with two seeds to a 12cm (4in) pot. It is a good idea to use peat pots to avoid root disturbance on planting out. Cover the pots with glass and paper or an opaque polythene bag and place them in a warm spot. Look at the seedlings every day, and as soon as they germinate remove the glass and paper and place them near the light. It may be necessary to shade the young seedlings from strong sunlight to prevent scorch. This is done by simply covering them with a piece of newspaper, which must be removed as soon as possible.

When the seedlings are growing well, remove the weakest. Keep them well watered until planting-out time.

The fertilizer in the compost should be sufficient until they are planted out, but, if they become pot-bound, they should be fed with a weak solution of general liquid fertilizer.

Planting

Further south, it should be possible to plant out under glass in mid-May, but in the north, it should be delayed until the end of the month or even early June. Outside planting should never be done before early June. There is no advantage in planting out too early. If conditions are too cold, the plants will become 'hard', and stop growing.

Set out the cloches about a week or so before the melons are to be planted so that the soil is warm at planting time. Before doing so, the ground should be prepared by deep digging, together with the incorporation of plenty of well-rotted farmyard manure or compost. This will help to retain moisture—an essential for successful growing.

When the cloches are set out, rake in about 120g per sq m (4oz per sq yd) of a tomato-base fertilizer.

Water the pots well before removing the plants and set them out in rows about 90cm (3ft) apart. If they are grown in frames, one plant at 90cm (3ft) intervals set in the middle of the frame can be expected to fill the allotted space.

Plant with the top of the rootball about 2.5cm (1in) above the ground. This will prevent water standing around the base of the stem and should avoid stem rot.

Planting out under cloches

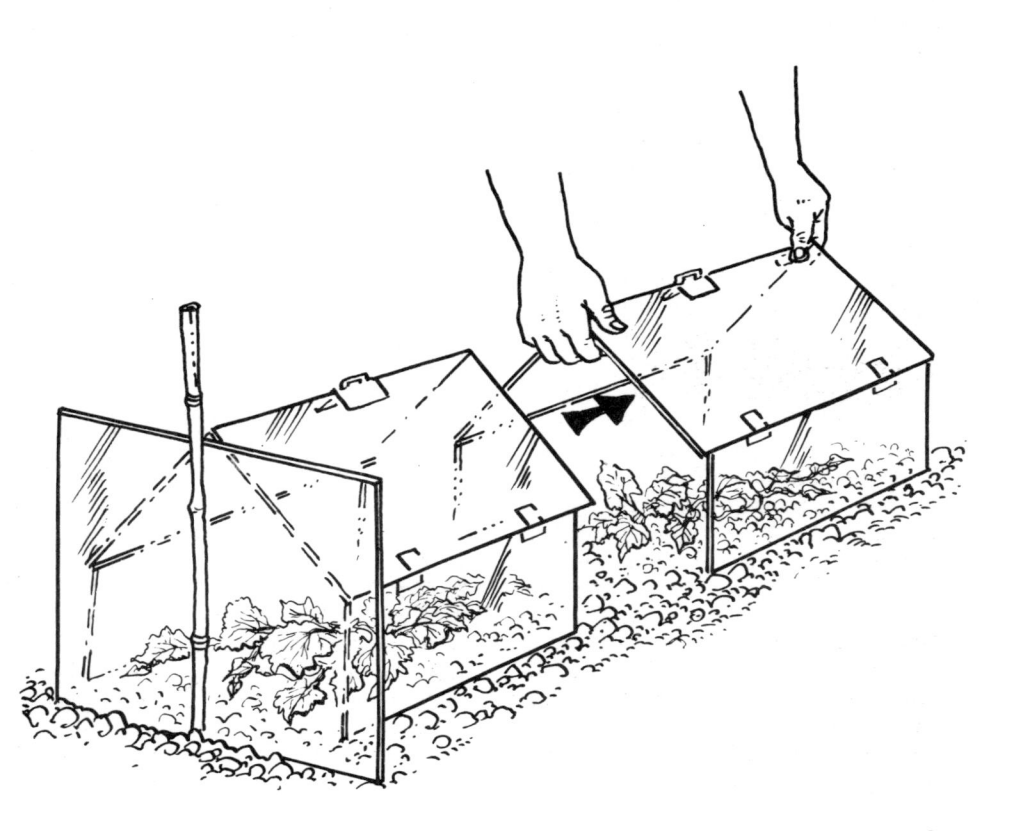

Melons

Cultivation
After planting, water the plants in well, and put the ends on the cloches or close the frame-lights completely.

After two or three days, depending on the weather, the frame lights can be opened a little during the day, or the cloches moved a little way apart. They must be closed up each evening, however. As the weather improves and the plants root into the soil and start to grow, ventilation can be increased.

Stop the plant after it has made five leaves, by pinching out the growing tip. This will induce the production of side-shoots.

Select four strong shoots and pinch out others that may form. Train the four shoots to the corners of the cold-frame or, if grown under cloches, as far apart as possible, two in each direction.

Allow three or four fruits to form on each side-shoot and then pinch out the tips. Spray the leaves regularly with clear water.

In warm weather the frames should be opened completely to allow pollinating insects to enter. When growing under cloches, remove one cloche completely at intervals.

With most hardy varieties, insect pollination should be sufficient, but if the weather is cold, they may not be very active. In this case, it is safer to pollinate by hand. This is done by removing a male flower, turning back the petals, and rubbing it into an open female flower. The female flower can be recognized by a tiny embryo fruit behind the petals.

The base fertilizer will be sufficient to keep the plants growing well until the first fruits set. When they do, they should be fed at weekly intervals with a proprietary tomato fertilizer according to the maker's instructions. Feeding should cease once the fruits start to ripen.

When the fruits begin to swell, keep them off the ground by placing a piece of slate or glass underneath them.

Harvesting
Once ripe, the fruit will give off a heavy aroma. If the blossom end is pressed, it should feel soft.

Pests and Diseases
Slugs can attack shoots, leaves and fruits. Protect them with slug pellets and by raising the fruit off the ground.

Greenfly can attack young shoots, sucking the sap and arresting growth. Control with resmethrin.

Red spider mite often attacks if the atmosphere is too dry. Spray leaves with water regularly, and if small, mottled patches appear, spray with dimethoate.

Mildew. This too is worse in dry conditions. Keep the atmosphere humid and, if the tell-tale white patches are seen, spray immediately with benomyl.

Footrot shows as a black, slimy area at the base of the stem. Keep the plants dry around the base of the stem and allow adequate ventilation.

Varieties

Charentais. A variety with excellent flavour that crops very early and is ideal for growing under cold glass or outside.

Ogen. Small, round fruits of fine flavour. Early and good for growing under cold glass.

Sweetheart. Rarely fails under cold glass or outside in warmer areas. It produces small juicy fruits of excellent flavour.

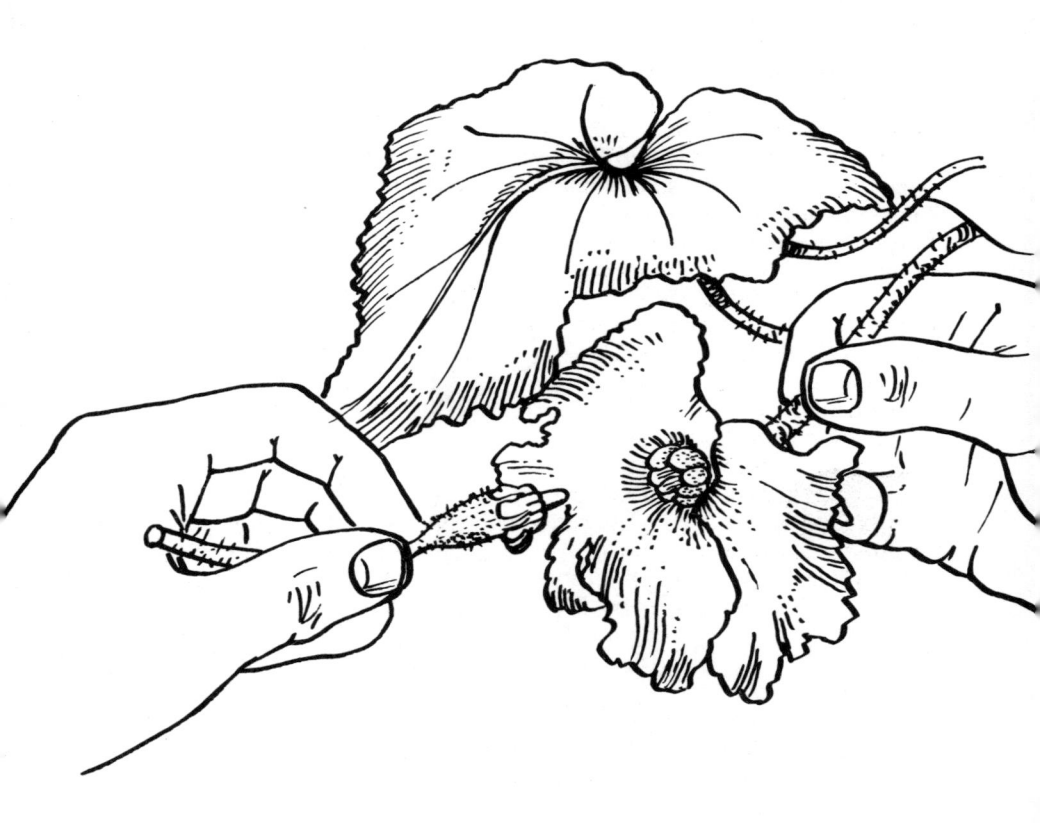

Pollinating by hand

Raspberries

Provided virus diseases can be kept at bay, raspberries are an excellent soft fruit to grow in the garden. They crop early—eighteen months after planting—are fairly soil tolerant, and produce a large amount of fruit. The fruits are superb fresh, they can be bottled or made into jam, and they freeze very well.

Preparation

Raspberries do best in a deep, well-drained soil that will hold plenty of moisture. They will, however, grow on most soils provided they are not badly drained or too limy. During the summer before planting, dig a trench two spades deep and about 75 cm (2ft 6in) wide, working a generous amount of well-rotted farmyard manure or compost into the bottom.

About two weeks before planting, rake in a dressing of bonemeal at 120g per sq m (4oz per sq yd).

Ideally, the trench will be running north to south and in the sun.

Planting

When buying canes, make sure that they carry a Ministry certificate of trueness to variety and freedom from virus disease. This is very important since there is no cure for virus diseases except to pull out the canes and burn them.

Prepare the ground for planting by digging a trench and working in well-rotted manure or compost

Never accept canes from a friend.

The best time to plant is early November when there is some warmth in the soil, but planting can continue right though until early April.

Set the canes 45cm (18in) apart and, if you plant more than one row, leave 2m (6ft) between rows.

Plant so that the top of the root is buried about 7.5cm (3in) and firm the plants in well with your heel.

A short time after planting, cut the canes back to leave about 23cm (9in) above the ground.

At about this time, it is a good idea to erect the supporting framework.

Support

Since the crop will be growing for several years, it is well worth while making a good, permanent job of the supports.

Set two stout poles at either end of the row. The poles should be of timber at least 7.5 × 7.5cm (3in × 3in) and it should ideally be 'tanalized' (pressure-treated with preservative). If not, soak the bottoms of the poles in a copper-based preservative (*not* creosote) for a day and a night before use. Use 2.2m (7ft)-long posts and sink them 45cm (18in) in the ground. It is well to strut the posts. On long rows, set intermediate posts at about 3m (10ft) apart.

Stout galvanized wires should be fixed to the posts 60cm (2ft), 1m (3ft 3in) and 1.5m (5ft) from the ground.

As the canes grow, they should be tied into these wires to prevent them from breaking.

Cultivation

The plants will not fruit in their first year, and flowers should be removed. This will establish strong canes for fruiting the following year.

Keep the rows weed-free either by very shallow hoeing or by the use of weedkillers such as paraquat (Weedol) or glyphosate (Tumbleweed). Hoeing must be done with care since the mass of surface roots made by raspberries are easily damaged. When the ground is clean and moist, it can be mulched with manure or compost in early May every year.

A dressing of sulphate of potash applied at 22g per sq m ($\frac{3}{4}$oz per sq yd) should be applied each March. Make sure that the canes never go short of water. After the first year, when the canes are fruiting, watering as the fruit is beginning to swell will increase the weight of the crop considerably.

Tying canes to a supporting framework

Raspberries

In the autumn after planting, remove any weak canes and any that are growing into the paths. The strong young canes should be tied in so that they are about 7.5–10cm (3–4in) apart.

Early in the second year, the canes should be 'tipped', cutting away a few inches to leave about 15cm (6in) above the top wire.

Apply the same feed and mulch as in the first year.

After picking the crop, the canes that have fruited should be cut right out to make room for the new canes that will bear fruit the following year. These should be tied in as before.

Autumn-fruiting Varieties

Autumn-fruiting varieties are grown in almost the same way. They will fruit in September or October. The canes are left unpruned on these until February, when they too should be cut back hard. The young canes will fruit in the same year.

Propagation

Raspberries are among the simplest fruits to propagate. Simply select a few strong-growing suckers that are encroaching on to the path where they are not wanted. These can be lifted with a quantity of root, and transplanted in rows. It cannot be stressed strongly enough, however, that it is folly to do this except from the healthiest material that has not been in the garden too long and has been regularly sprayed against greenfly—the carriers of virus diseases.

Pests and Diseases

Aphids are a common pest in the garden and the canes should be sprayed as soon as the pest is seen. They suck the sap, causing distortion and retarded growth, but worse, they carry virus diseases. As a preventative, spray with dimethoate at fortnightly intervals from late April. It is also a good idea to spray the dormant canes with tar-oil in December or January to kill overwintering eggs.

Raspberry beetle lays its eggs in the blossoms, and the grubs, when hatched, burrow into the fruit. It is essential to spray whether the pest is seen or not, since it is almost bound to attack and will not be seen until after the fruit is picked.

Removing suckers for propagation

Spray with malathion or fenitrothion when the first pink fruits are seen.

Raspberry cane midge causes small mottled areas on canes. Control by spraying with fenitrothion in early May and again two weeks later, though this second spraying should coincide with that against raspberry beetle.

Spur blight is seen as purple blotches around the spurs. They subsequently turn silver. Buds and shoots will later die back. Control by spraying with dichlorfluanid when the young canes are 5cm (2in) high and continue at fortnightly intervals until the first pink fruit. Then spray with benomyl. After picking, revert back to dichlorfluanid.

Cane spot shows as small purple blotches which later become grey, elliptical patches. It affects canes, leaves and fruits. It is controlled in the same way as cane spot.

Virus diseases cause yellow mottling and blotching on the leaves, and the canes lose vigour. Pull affected canes out and burn them. Grow new canes in another part of the garden.

Varieties

Malling Promise. A heavy, regular cropper, this variety has become very popular. The fruit is well flavoured and medium to large. It is fairly resistant to virus diseases but has the disadvantages of being susceptible to grey mould in wet weather, and of producing an over-abundance of young canes, especially early in its life.

Malling Jewel. A variety that can be generally recommended. Its fruit, which ripens in July, is large, firm and juicy and has an excellent flavour. Though it is a vigorous grower, it does not produce an embarrassing number of canes, and it is tolerant of virus disease. It is somewhat susceptible to cane blight.

Malling Delight. An early to mid-season variety recently introduced. It produces large fruits and crops well. Fruit is of good flavour but perhaps a little pale. It is resistant to aphids and therefore should miss virus infection.

Malling Admiral. A mid-season to late variety cropping in August. The fruit is large, firm and well flavoured. It has some resistance to virus.

Glen Clover. This variety has a very long season and crops heavily. Its fruit is of moderate flavour and medium size but excellent for jam making and freezing. It is susceptible to virus.

Autumn-fruiting varieties
September. This variety will crop in August or September. It produces firm, medium-sized berries of fair flavour. It is not very vigorous, so canes should be planted slightly closer.

Zeva. Fruits from July to November, producing large, firm fruit of good flavour. It is moderately vigorous and heavy cropping.

Red and White Currants

These are grown in a completely different way from blackcurrants and can take up much less space. Grown as cordons, they are ideal along one side of the fruit cage, or they can be trained on walls and fences.

Planting
Like all soft fruit, they will benefit from the addition of farmyard manure, but they are less demanding than blackcurrants.

There is no certification scheme for red or white currants, and they should therefore be bought from a reputable nurseryman who will have kept them free from aphid attack. Two-year-old bushes should be bought. They should be planted in the dormant season as soon after the end of October as possible. If they are to be grown on as bushes, plant them 1.5m (5ft) apart with 2m (6ft) between the rows. If grown as single cordons, plant them 38cm (15in) apart.

Red and white currants, unlike blackcurrants, are grown on a short stem or 'leg', so they should be planted only very slightly deeper than at the nursery. After planting, cut the branches back by about half, cutting to an outward-facing bud.

Cultivation
Keep weeds down by shallow hoeing or weedkiller treatment. Mulch with well-rotted manure or compost in late spring. In February, feed the bushes with sulphate of potash at 22g per sq m ($\frac{3}{4}$oz per sq yd). They are very prone to potash deficiency, so this feed should not be neglected.

Sometimes suckers will appear at the base, and these should be torn out in the same way as with roses. In exposed places, the young shoots are likely to be blown out, so in this case, young bushes should be caned and the branches important to the shape of the bush tied in.

Pruning
Winter and summer pruning should be carried out. In the first winter after planting, if the plants are being grown as bushes, the leading shoots should be cut back by about half, and the laterals to within two buds of the base. The amount of pruning will depend upon the amount of growth the bush has made. If growth is weak, cut the shoots back harder.

Bushes must be planted at the correct depth

In the summer, at the end of June, shorten the laterals to five leaves but do not touch the leaders. In later years, the year's growth should be cut back to about 2.5cm (1in) in the winter, and the laterals shortened as before. Older branches must also be removed from time to time to make room for new young growths.

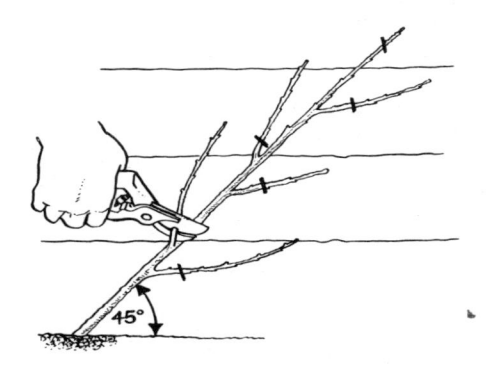

Pruning cordons

If the plants are being grown as cordons, they are normally planted at 45° to the ground, and must be trained against a fence or wall or on a post and wire structure.

In the winter, the leading shoots are cut back to about one third, side shoots to about 13mm ($\frac{1}{2}$in).

At the end of June, the side shoots are reduced to five leaves, but the leading shoot is left alone. When the plant has reached the required height, the leading shoot is treated in the same way as the side shoots, reducing it to five leaves.

Propagation
Red and white currants are propagated in the same way as gooseberries.

Pests and Diseases
Birds are undoubtedly the worst pest of currants. Plants can be sprayed with bird repellent, but this is not really very effective. For good control, net the bushes.

Aphids, the scourge of all soft fruits, will attack currants as well. Control with dimethoate when the first flowers open.

Coral spot causes red spots on old wood. It can cause die-back of younger wood if it spreads. Cut it out and paint the wound with pruning compound.

Leaf spot causes dark spots on the leaves, later spreading to form large blotches. Control as for blackcurrants.

Varieties

Redcurrants
Jonkheer van Tets. A very early variety. It produces very large berries of medium flavour for dessert or cooking. It flowers early, so it may not escape frosts.
Laxton's No 1. A popular early variety that is heavy yielding and reliable. It produces medium to large fruit of fine quality.
Red Lake. A mid-season variety that produces heavy crops of large, juicy berries, excellent for jelly.

White currants
White Versailles is the best garden variety. It produces long bunches of large berries, of excellent flavour.

Rhubarb

Rhubarb is easy to grow and very prolific. It is subject to few pests and diseases and is therefore a very good plant to include in the soft-fruit garden.

Strictly speaking, of course, rhubarb is not a fruit, since it is the stems that are used for cooking. As it is always used for fruit dishes, however, it is included here.

Preparation

Rhubarb will not grow well in dry soils. On the other hand, wet, waterlogged soil tends to cause rotting of the crowns, so good drainage must be provided. Since it will be grown in the same place for several years, it is worth while preparing the soil well before planting.

The soil should be deeply dug, and generous quantities of organic matter in the form of well-rotted manure or compost should be incorporated. The same materials should also be used as a mulch after planting and every year in the autumn.

When digging, ensure that all perennial weeds are removed and annual weeds buried.

Planting

Though rhubarb can be raised from seed, it is a long-winded job. It is better to buy some crowns from a garden centre, so that cropping can start in the second year.

The best time to plant is in October or November, though crowns can be planted in February or March. They can also be bought from garden centres in containers, and planting then can take place at any time.

Plant the sets 1m (3ft) square, so that the top of the crown is no more than about 5cm (2in) below the surface. After planting, firm the sets in well with your heel and spread a little well-rotted manure over the site.

Cultivation

There is really very little to do in the first year. The crowns will produce stems and large leaves that will quickly swamp all competition from weeds. It may be necessary in the early stages, however, to hoe lightly around the crowns. Above all, make sure that the plants never go short of water. In really dry weather, it may be necessary to water by hand.

Do not pull any stems in the first year, but keep a sharp lookout for flower heads. These will only weaken the plants, and should be removed as soon as they are seen.

In the early spring—February or March—the plants will benefit from a top-dressing of a general fertilizer. Apply Growmore around the crowns at a rate of about 120g per sq m (4oz per sq yd).

In the second year, some shoots can be pulled, but only a few. The plants will be productive for a further five years, so it is important to allow them to build up strength in the first two years. Never pull all the shoots, even in succeeding years, but leave enough for the plant to continue to maintain strength. Always pull the stems, twisting a little at the same time. Never cut them.

In the autumn, when the top growth has died down, clear away all the old litter from around the plants and mulch with a thick layer of strawy manure. This will protect the young crowns and produce the earliest crop. The mulch should not be allowed to settle down, but should be shaken up occasionally to allow a free flow of air.

Forcing
Apart from the gentle forcing mentioned above, rhubarb can be forced to give very early crops. This can be done outside, in the greenhouse or even in the kitchen.

It must be borne in mind that forcing will greatly weaken the crowns, so the normal practice is to force only those crowns that are due to be discarded after having served their seven-year term. Gentle forcing outside, however, can be done without too much harm, though it is wise to force a different crown each year.

The principle of forcing is simply to provide more warmth and to keep the stems in darkness.

Outside, the crowns are covered in January with a bucket, a large earthenware pot or a strong wooden box. The covering is then buried in straw, or, ideally, strawy horse manure for extra warmth. The sticks will be ready for pulling about three weeks before the normal time.

For forcing indoors, choose strong roots of three- or four-year-old plants. Dig them up in the autumn when the foliage has died down. If they are subjected to a cold spell outside after lifting, they will grow faster when brought into the warmth.

Box them up in a little compost and water them. Put the boxes in a dark place in a temperature of 10–13°C (50–55°F).

Water the roots occasionally to keep the compost just moist. The stems will be ready for pulling in about six weeks.

It is a good idea to keep two or three boxes on the go, boxing up the roots at different times to maintain a succession until the outside plants are cropping. After forcing, the crowns will be exhausted and should be thrown away.

Propagation
Though it can be raised from seed sown outside in April, rhubarb is better propagated vegetatively.

This is done simply by digging up a root, and dividing it with a sharp spade or a knife. Make sure that each division has a bud and a piece of root.

Pests and Diseases
About the only disease likely to affect rhubarb is crown rot, for which there is no known control. It causes a rotting of the crown and the base of the leaves. Infected crowns should be dug up and burnt.

Varieties
Timperley Early. This is a very early variety and good for forcing. In warm areas it will be cropping in February though normally it will yield its first stems in March.
Hawke's Champagne. Another early variety often used for forcing.
The Sutton. A good maincrop variety that has a fine flavour and rarely runs to seed.
Victoria. A well-known maincrop variety.

Strawberries

Strawberries are an exception to the soft-fruit rule. They are grown in quite a different way from other types.

They take up a fair bit of room in the fruit garden and are perhaps not as heavy-cropping per square metre as other soft fruits. But the quality of fresh, garden-grown strawberries is such that they cannot be excluded.

New varieties and new techniques have increased their cropping potential and reduced the problems of disease. Many gardeners now grow them as annuals, for example, putting down a single row in the vegetable plot rather than making a permanent bed. This way is no more expensive and will increase yields.

Preparation

Strawberries will crop quite well in most soils. If they are to be grown as annuals, they are better in soil that has not been overmanured or fertilized, but if a semi-permanent bed is to be made, it is wise to manure the ground before planting. An over-rich soil will produce too much leaf growth at the expense of fruit. They are best grown in full sun.

Planting

The best time to plant is August or early September. If they are left any later, they should not be allowed to crop the following summer.

Make sure that plants are certified free from virus infection. This

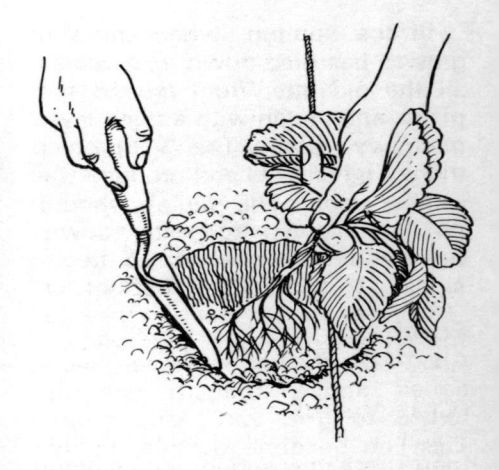

When planting, do not bury the crown

is very important since strawberries are very prone to virus diseases which will live in the soil for some time.

It may be difficult to find plants that have been Ministry certified as early as August, but if they are bought from a reputable nurseryman and have been entered for certification, they are a safe bet.

If you intend to grow the plants in beds, make them about 2.3m (7ft 6in) wide. This will allow three rows, and you will be able to reach comfortably to the middle of the row.

About a week before planting, rake into the surface a dressing of bone-meal at 90g per sq m (3oz per sq yd) and sulphate of potash at 15g per sq m ($\frac{1}{2}$oz per sq yd). At the same time, rake the soil to a fine, level tilth.

Mark out the first row 38cm (15in) from the edge of the bed. The remaining rows will be 76cm (30in) apart, and plants should be 38cm (15in) apart in the rows.

When planting, it is essential to ensure that the crown of the plant is not buried. When the hole is dug, place a cane across it to make sure that the planting depth is right. Spread out the roots in the hole and refill with soil, working it between the roots with your fingers. Firm the soil well around the roots and water the plants in. They may look a little sad for a day or two, but they will soon recover.

Cultivation

Plants will need very little attention now until the spring. Keep down weeds by hand pulling. If you do use a hoe, make sure that the cultivations are very shallow to avoid damaging the roots.

The danger period is when the plants start flowering. They are prone to damage by late frosts, and if this happens, they will not fruit, of course. If frost threatens, cover the plants with a little straw, or even newspaper.

If you do cover the plants at night, make sure they are uncovered the following morning. Light frost damage can be avoided by spraying the flowers with water in the late evening.

Remove the runners as they appear unless they are needed for propagation purposes, and never allow the plants to go short of water. Heavy watering should be avoided, however, in the early stages. Too much water early on will induce a lot of leaf growth, but will not increase the crop. The time to apply water (unless, of course, the plants are under stress), is when the fruits are swelling. This will greatly increase the size of the berries and the weight of the crop. When the fruit begins to ripen fast, watering should be stopped.

As the fruit starts to swell, the plants can be mulched with straw. This will prevent the attentions of slugs, and will protect the fruit from dirt. Alternatively, use black polythene mulching strip, or special 'whalehide' strawberry mats. It

Strawberry mats provide protection

is important not to apply the mulch too early, since this could result in frost damage. Mulching will insulate the plants from the warmth rising from the soil at night, and those few extra degrees could make all the difference.

Once the fruit starts to ripen, it *must* be protected against birds. They will almost certainly try to get to the ripe fruit before you do, so a covering with netting is essential.

Strawberries

Harvesting

Harvest the fruit with care—it is easily damaged. To avoid bruising, take the calyx and a little bit of stalk as well. Pick over the beds regularly to encourage continuous cropping.

After harvesting, cut the plants back quite hard. All the old leaves can be removed, but the crowns and the young leaves must be left intact.

Remove all the old leaves, together with the mulch, and pull out any weeds there may be. At the same time, remove any runners that are not wanted and burn the lot. After thoroughly cleaning the beds, give them a top-dressing of sulphate of potash at 15g per sq m ($\frac{1}{2}$oz per sq yd). Apply a general fertilizer only if the plants are not growing well. Strawberries will search for the food they need and should not be overfed. The plants will decrease in vigour over the years, and generally the first crop will be the heaviest. This is why some gardeners grow them as annuals.

Beds should therefore be replaced after three or, under ideal conditions, four years. New plants can be provided by taking runners from existing plants, but as a precaution against disease, new plants should be bought in periodically.

Alternative Methods of Growing

Growing under Cloches

An earlier crop can be produced by using cloches, either the 'barn' type or polythene tunnels. Early strawberries are particularly welcome because at that time shop prices tend to be high. The plants are set out at the same time as those outside, but distances will depend upon the size of the cloches. Probably a single row with the plants 23cm (9in) apart will be sufficient.

Strawberries need a period of cold to initiate fruit buds, so don't put the cloches over the plants too soon. This will only result in a lot of lush growth and very little fruit. Covering depends upon location, but generally February is about right for the Midlands and south, and a little earlier farther north.

When the plants begin to flower, they will need some ventilation. Move barn cloches apart and lift the sides of polythene tunnels. The spaces should be enough to allow insects to enter, but to exclude birds. Do not remove the end glasses of barn cloches, since this will create a 'wind-tunnel' effect.

After harvesting, remove the cloches and treat the plants in the normal way.

Growing as Annuals

Because the heaviest crop comes in the first year after a late summer planting, some gardeners prefer to grow strawberries as annuals.

The plants are grown in rows, rather than in beds, to allow room for runner production. Only one runner per plant is allowed to develop, and this is pegged down into the ground to root. When the plants have finished cropping, they are removed and burnt, and the row propagated by the runners is used for next year's crop.

To avoid the build-up of disease, it is advisable to replace the plants with fresh, virus-free stock every three or four years.

Growing through Polythene

To avoid excessive weed growth and to prevent runners rooting, strawberries are sometimes grown through polythene mulching sheet. Make up a bed so that the surface is slightly convex, to allow water to run off. Bury the black polythene in shallow slit-trenches either side of the bed. The plants are then set in the normal way, through slits cut in the sheet.

This method has the disadvantage that watering is difficult, and it is only really practicable if a trickle irrigation line can be laid in the centre of the bed beneath the polythene.

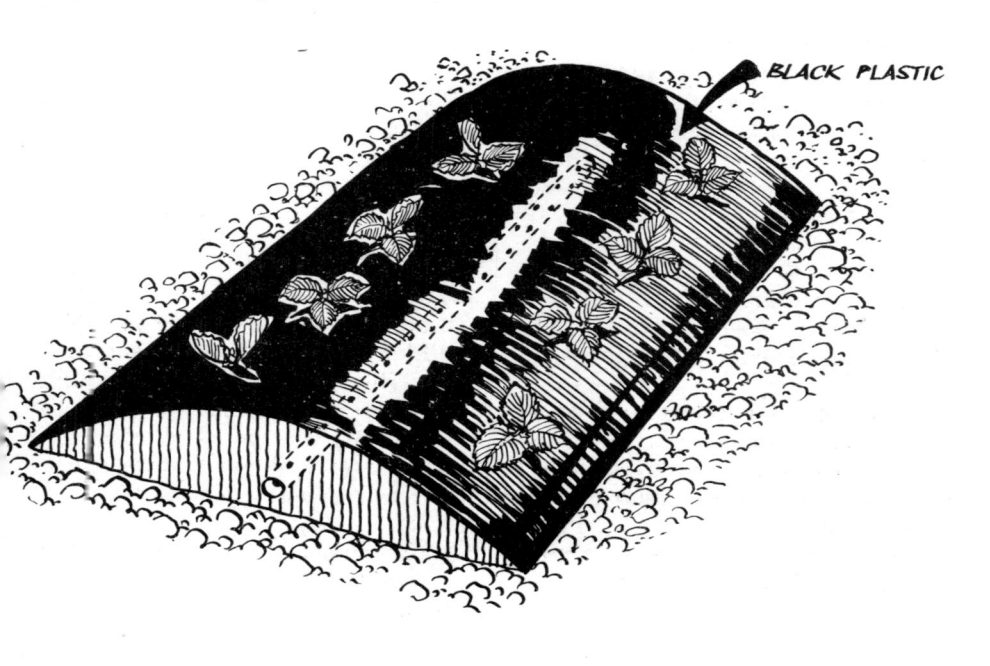

Growing strawberry plants through a polythene mulching sheet

Strawberries

Growing in Tubs and Towers

Strawberries make excellent subjects for growing in barrels or tubs. Special barrels and pots with holes in the sides can be obtained, and there are also some attractive 'tower-pots' that look good on the patio, or which can be used in the greenhouse. Containers enable the gardener with limited space to grow plants in a very small area.

Fill the containers with a good open compost—equal parts loam, peat and sharp sand is ideal—and grow them in the normal way. Remember though, that extra water and fertilizer may be needed since containers tend to dry out more quickly than the open ground.

Propagation

New plants are raised quite easily by simply pegging runners into the soil. Ideally, unless the plants are being grown as annuals as previously described, a few plants should be grown specifically to produce runners. Use plants that have previously given good yields of fruit, and plant them in another part of the garden away from the fruiting plants.

All the flowers should be removed from these plants. This will channel all the strength of the plant into leaf and runner production and the resulting plants will be stronger.

In early July, the runners should be pegged into the soil with wire

Pegging runners into pots of compost set into the soil

'staples'. Better still, to avoid subsequent root disturbance, peg them into pots of compost, sunk to their rims in the soil.

The runners will have rooted by August and can be replanted.

Pests and Diseases

Greenfly. Spraying against greenfly *must* be carried out to avoid the spread of virus diseases. Spray just before flowering with dimethoate, and if the attack is severe, again after harvesting.

Red spider mite produces small, whitish areas which later turn bronze. Spray with dicofol just before flowering.

Strawberry mite causes leaves to become rough and wrinkled and kills young leaves. Spray with dicofol after cropping.

Strawberry blossom weevil partially severs the bud stem. Control with malathion or fenitrothin as soon as the symptoms are seen.

Botrytis is a common fungus disease. It can be easily recognized as a grey mould on fruit and stems. Control by spraying four times with dichlorfluanid, thiram or captan—first when the first erect buds show, secondly when the first flowers open, thirdly at petal fall and finally after cropping.

Virus diseases cannot be controlled. Buy certified stock and remove and burn any that become infected.

Varieties

Cambridge Favourite is the most popular commercial variety and the one you would buy in the shops. It is very heavy cropping, but the flavour is only fair. It is tolerant of virus diseases, but the fruit is subject to grey mould in wet weather.

Cambridge Vigour produces medium to large berries in the first year, but they tend to be smaller in later years. It is therefore a good variety to grow as an annual. Flavour is good.

Grandee has very large berries of excellent flavour, and is somewhat irregular in shape.

Royal Sovereign is perhaps the best-flavoured of all varieties. The berries are large and freely produced. It is, however, very susceptible to virus diseases.

Tamella is a new variety and therefore not entirely predictable. It often produces a second crop in the autumn especially if cloched in the spring. The fruit is medium-sized with a good flavour.

Remontant Varieties

Remontant varieties will crop after the summer varieties have finished. They are grown in the same way as normal varieties, except that the first flowers should be removed until the end of May. They are more demanding of feed and water than other types. Gento and Rabunda are recommended varieties.

Illustrated by Barry Gurbutt

Hamilton, Geoff
Growing soft fruits.—(Penny pinchers).
G1. Fruit-culture
I. Title II. Series
I634'.7   SB381
ISBN 0-7153-7903-8

Text and illustrations
© David & Charles Ltd 1980

All rights reserved. No part of this publication may be reproduced, stored in a retrieval system, or transmitted, in any form or by any means, electronic, mechanical, photocopying, recording or otherwise, without the prior permission of David & Charles (Publishers) Limited

Printed in Great Britain
by A. Wheaton & Co., Ltd
for David & Charles (Publishers) Limited
Brunel House   Newton Abbot   Devon

Published in the United States of America
by David & Charles Inc
North Pomfret   Vermont 05053   USA